MODERN HPLC FOR PRACTICING SCIENTISTS

Michael W. Dong

Synomics Pharmaceutical Services, LLC
Wareham, Massachusetts

WILEY-INTERSCIENCE

A JOHN WILEY & SONS, INC., PUBLICATION

For general information on our other products and services or for technical support, please contact our Customer Care Department within the United States at (800) 762-2974, outside the United States at (317) 572-3993 or fax (317) 572-4002.

Wiley also publishes its books in a variety of electronic formats. Some content that appears in print may not be available in electronic formats. For more information about Wiley products, visit our web site at www.wiley.com.

Library of Congress Cataloging-in-Publication Data:

Dong, M. W.
    Modern HPLC for practicing scientists / by Michael W. Dong.
       p. cm.
    Includes bibliographical references and index.
    ISBN-13: 978-0-471-72789-7
    ISBN-10: 0-471-72789-X
    1. High performance liquid chromatography. 2. Drugs—Analysis. I. Title.
    RS189.5.H54D66 2006
    615′.19—dc22

                                2005057463

Printed in the United States of America.

10 9 8 7 6 5 4 3 2 1

CONTENTS

PREFACE

The idea for writing this basic HPLC book was probably born in the New York City subway system while I was a graduate student in the 1970s. Amidst the rumbling noise of the subway, I was reading "the green book"—*Basic Gas Chromatography* by McNair and Bonnelli—and was immediately impressed with its simplicity and clarity. In the summer of 2004, I had just completed the editing of *Handbook of Pharmaceutical Analysis by HPLC* with Elsevier/Academic Press, and was toying with the idea of starting a book project on Fast LC and high-throughput screening. Several phone conversations with Heather Bergman, my editor at Wiley, convinced me that an updated book on modern HPLC, modeled after "the green book," would have more of an impact.

This book was written with a sense of urgency during weekends and weekday evenings ... through snow storms, plane trips, allergy seasons, company restructuring, and job changes. The first draft was ready in only 10 months because I was able to draw many examples from my previous publications and from my short course materials for advanced HPLC in pharmaceutical analysis given at national meetings. I am not a fast writer, but rather a methodical one who revised each chapter many times before seeking review advice from my friends and colleagues. My goal was to provide the reader with an updated view of the concepts and practices of modern HPLC, illustrated with many figures and case studies. My intended audience was the practicing scientist—to provide them with a review of the basics as well as best practices, applications, and trends of this fast-evolving technique. Note that this basic book for practitioners was written at both an introductory and intermediate level. I am also targeting the pharmaceutical analysts who constitute a significant fraction of all HPLC users. My focus was biased towards reversed-phase LC and pharmaceutical analysis. The scope of this book does not allow anything more than a cursory mention of the other applications.

Writing a book as a sole author was a labor of love, punctuated with flashes of inspiration and moments of despair. It would have been a lonely journey without the encouragement and support of my colleagues and friends. First and foremost, I would like to acknowledge the professionalism of my editor at John Wiley, Heather Bergman, whose enthusiasm and support made this a happy project. I also owe much to my reviewers, including the 10 reviewers of the book proposal, and particularly to those whose patience I tested by asking them to preview multiple chapters. They have given me many insights and valuable advice. The list of reviewers is long:

Prof. David Locke of City University of New York (my graduate advisor); Prof. Harold McNair of Virginia Tech, whose "green book" provided me with a model; Prof. Jim Stuart of University of Connecticut; Drs. Lloyd Snyder and John Dolan of LC Resources; Drs. Raphael Ornaf, Cathy Davidson, and Danlin Wu, and Joe Grills, Leon Zhou, Sung Ha, and Larry Wilson of Purdue Pharma; Dr. Ron Kong of Synaptic; Drs. Uwe Neue, Diane Diehl, and Michael Swartz of Waters Corporation, Wilhad Reuter of PerkinElmer; Drs. Bill Barbers and Thomas Waeghe of Agilent Technologies; Drs. Krishna Kallary and Michael McGinley of Phenomenex; Dr. Tim Wehr of BioRad; John Martin and Bill Campbell of Supelco; Dr. Andy Alpert of PolyLC; Margie Dix of Springborn-Smithers Laboratories; Dr. Linda Ng of FDA, CDER; and Ursula Caterbone of MacMod.

Finally, I acknowledge the support and the unfailing patience of my wife, Cynthia, and my daughter, Melissa, for putting up with my long periods of distraction when I struggled for better ways for putting ideas on paper. To them, I pledge more quality time to come after 2006.

Norwalk, Connecticut   MICHAEL W. DONG

1

INTRODUCTION

1.1 INTRODUCTION

1.1.1 Scope

High-performance liquid chromatography (HPLC) is a versatile analytical technology widely used for the analysis of pharmaceuticals, biomolecules,

Modern HPLC for Practicing Scientists, by Michael W. Dong
Copyright © 2006 John Wiley & Sons, Inc.

polymers, and many organic and ionic compounds. There is no shortage of excellent books on chromatography[1,2] and on HPLC,[3–9] though many are outdated and others cover academic theories or specialized topics. This book strives to be a concise text that capsulizes the essence of HPLC fundamentals, applications, and developments. It describes basic theories and terminologies for the novice and reviews relevant concepts, best practices, and modern trends for the experienced practitioner. While broad in scope, this book focuses on reversed-phase HPLC (the most common separation mode) and pharmaceutical applications (the largest user segment). Information is presented in a straightforward manner and illustrated with an abundance of diagrams, chromatograms, tables, and case studies and supported with selected key references or web resources.

Most importantly, this book was written as an updated reference guide for busy laboratory analysts and researchers. Topics covered include HPLC operation, method development, maintenance/troubleshooting, and regulatory aspects. This book can serve as a supplementary text for students pursuing a career in analytical chemistry. A reader with a science degree and a basic understanding of chemistry is assumed.

This book offers the following benefits:

- A broad-scope overview of basic principles, instrumentation, and applications.
- A concise review of concepts and trends relevant to modern practice.
- A summary update of best practices in HPLC operation, method development, maintenance, troubleshooting, and regulatory compliance.
- A summary review of modern trends in HPLC, including quick-turnaround and "greener" methods.

1.1.2   What Is HPLC?

Liquid chromatography (LC) is a physical separation technique conducted in the liquid phase. A sample is separated into its constituent components (or analytes) by distributing between the mobile phase (a flowing liquid) and a stationary phase (sorbents packed inside a column). For example, the flowing liquid can be an organic solvent such as hexane and the stationary phase can be porous silica particles packed in a column. HPLC is a modern form of LC that uses small-particle columns through which the mobile phase is pumped at high pressure.

Figure 1.1a is a schematic of the chromatographic process, where a mixture of analytes A and B are separated into two distinct bands as they migrate down the column filled with packing (stationary phase). Figure 1.1b is a representation of the dynamic partitioning process of the analytes between the flowing liquid and a spherical packing particle. Note that the movement of component B is retarded in the column because each B molecule has stronger affinity for

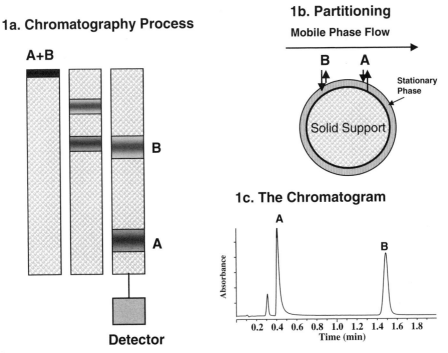

1a. Chromatography Process

1b. Partitioning

Mobile Phase Flow

1c. The Chromatogram

Detector

Figure 1.1. (a) Schematic of the chromatographic process showing the migration of two bands of components down a column. (b) Microscopic representation of the partitioning process of analyte molecules A and B into the stationary phase bonded to a spherical solid support. (c) A chromatogram plotting the signal from a UV detector displays the elution of components A and B.

the stationary phase than the A molecule. An in-line detector monitors the concentration of each separated component band in the effluent and generates a trace called the "chromatogram," shown in Figure 1.1c.

1.1.3 A Brief History

Classical LC, the term *chromatography* meaning "color writing," was first discovered by Mikhail Tswett, a Russian botanist who separated plant pigments on chalk ($CaCO_3$) packed in glass columns in 1903.[10] Since the 1930s, chemists used gravity-fed silica columns to purify organic materials and ion-exchange resin columns to separate ionic compounds and radionuclides. The invention of gas chromatography (GC) by British chemists A.J.P. Martin and co-workers in 1952, and its successful applications, provided both the theoretical foundation and the incentive for the development of LC. In the late 1960s, LC turned "high performance" with the use of small-particle columns that required high-pressure pumps. The first generation of high-performance liquid chromatographs was developed by researchers in the 1960s, including Horvath,

2a   2b

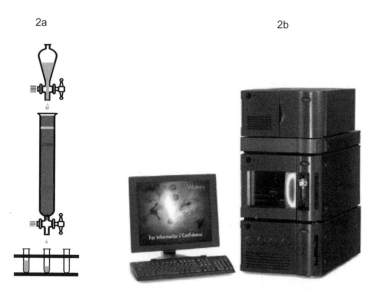

Figure 1.2. (a) The traditional technique of low-pressure liquid chromatography using a glass column and gravity-fed solvent with manual fraction collection. (b) A modern automated HPLC instrument (Waters Acquity UPLC system) capable of very high efficiency and pressure up to 15,000 psi.

Kirkland, and Huber. Commercial development of in-line detectors and reliable injectors allowed HPLC to become a sensitive and quantitative technique leading to an explosive growth of applications.[10] In the 1980s, the versatility and precision of HPLC rendered it virtually indispensable in pharmaceuticals as well as other diverse industries. The annual worldwide sales of HPLC systems and accessories approached three billion US$ in 2002.[11] Today, HPLC continues to evolve rapidly toward higher speed, efficiency, and sensitivity, driven by the emerging needs of life sciences and pharmaceutical applications. Figure 1.2a depicts the classical technique of LC with a glass column that is packed with coarse adsorbents and gravity fed with solvents. Fractions of the eluent containing separated components are collected manually. This is contrasted with the latest computer-controlled HPLC, depicted in Figure 1.2b, operated at high pressure and capable of very high efficiency.

1.1.4   Advantages and Limitations

Table 1.1 highlights the advantages and limitations of HPLC. HPLC is a premier separation technique capable of multicomponent analysis of real-life samples and complex mixtures. Few techniques can match its versatility and precision of <0.5% relative standard deviation (RSD). HPLC is highly automated, using sophisticated autosamplers and data systems for unattended analysis and report generation. A host of highly sensitive and specific detec-

Table 1.1. Advantages and Limitations of HPLC

Advantages
  • Rapid and precise quantitative analysis
  • Automated operation
  • High-sensitivity detection
  • Quantitative sample recovery
  • Amenable to diverse samples
Limitations
  • No universal detector
  • Less separation efficiency than capillary GC
  • More difficult for novices

tors extend detection limits to nanogram, picogram, and even femtogram levels. As a preparative technique, it provides quantitative recovery of many labile components in milligram to kilogram quantities. Most importantly, HPLC is amenable to 60% to 80% of all existing compounds, as compared with about 15% for GC.[3,4]

HPLC suffers from several well-known disadvantages or perceived limitations. First, there is no universal detector, such as the equivalence of flame ionization detector in GC, so detection is more problematic if the analyte does not absorb UV hays or cannot be easily ionized for mass spectrometric detection. Second, separation efficiency is substantially less than that of capillary GC, thus, the analysis of complex mixtures is more difficult. Finally, HPLC has many operating parameters and is more difficult for a novice. As shown in later chapters, these limitations have been largely minimized through instrumental and column developments.

1.2 MODES OF HPLC

In this section, the four major separation modes of HPLC are introduced and illustrated with application examples, each labeled with the pertinent parameters: column (stationary phase), mobile phase, flow rate, detector, and sample information. These terminologies will be elaborated later.

1.2.1 Normal-Phase Chromatography (NPC)

Also known as liquid-solid chromatography or adsorption chromatography, NPC is the traditional separation mode based on adsorption/desorption of the analyte onto a polar stationary phase (typically silica or alumina).[3–5] Figure 1.3a shows a schematic diagram of part of a porous silica particle with silanol groups (Si-OH) residing at the surface and inside its pores. Polar analytes migrate slowly through the column due to strong interactions with the silanol groups. Figure 1.4 shows a chromatogram of four vitamin E isomers in a palm

3a   3b

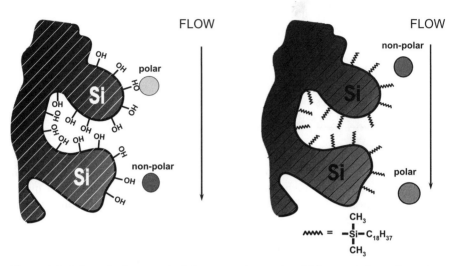

Figure 1.3. Schematic diagrams depicting separation modes of (a) normal-phase chromatography (NPC) and (b) reversed-phase chromatography (RPC).

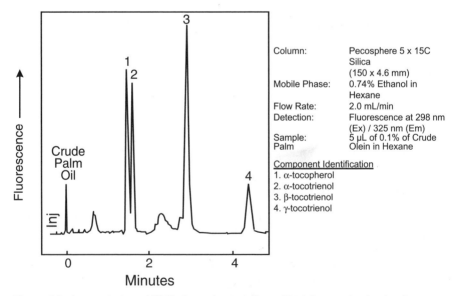

Column:	Pecosphere 5 x 15C Silica (150 x 4.6 mm)
Mobile Phase:	0.74% Ethanol in Hexane
Flow Rate:	2.0 mL/min
Detection:	Fluorescence at 298 nm (Ex) / 325 nm (Em)
Sample:	5 µL of 0.1% of Crude Palm Olein in Hexane

Component Identification
1. α-tocopherol
2. α-tocotrienol
3. β-tocotrienol
4. γ-tocotrienol

Figure 1.4. A normal-phase HPLC chromatogram of a palm olein sample showing the separation of various isomers of vitamin E. Chromatogram courtesy of PerkinElmer.

olein sample using a nonpolar mobile phase of hexane modified with a small amount of ethanol. It is believed that a surface layer of water reduces the activity of the silanol groups and yields more symmetrical peaks.[3] NPC is particularly useful for the separation of nonpolar compounds and isomers, as well as for the fractionation of complex samples by functional groups or for sample clean-up. One major disadvantage of this mode is the easy contamination of the polar surfaces by sample components. This problem is partly reduced by bonding polar functional groups such as amino- or cyano-moiety to the silanol groups.

1.2.2 Reversed-Phase Chromatography (RPC)

The separation is based on analytes' partition coefficients between a polar mobile phase and a hydrophobic (nonpolar) stationary phase. The earliest stationary phases were solid particles coated with nonpolar liquids. These were quickly replaced by more permanently bonding hydrophobic groups, such as octadecyl (C18) bonded groups, on silica support. A simplified view of RPC is shown in Figure 1.3b, where polar analytes elute first while nonpolar analytes interact more strongly with the hydrophobic C18 groups that form a "liquid-like" layer around the solid silica support. This elution order of "polar first and nonpolar last" is the reverse of that observed in NPC, and thus the term "reversed-phase chromatography." RPC typically uses a polar mobile phase such as a mixture of methanol or acetonitrile with water. The mechanism of separation is primarily attributed to solvophobic or hydrophobic interaction.[12,13] Figure 1.5 shows the separation of three organic components. Note that uracil, the most polar component and the most soluble compound in the mobile phase, elutes first. *t*-Butylbenzene elutes much later due to increased hydrophobic interaction with the stationary phase. RPC is the most popular HPLC mode and is used in more than 70% of all HPLC analyses.[3,4] It is suitable for the analysis of polar (water-soluble), medium-polarity, and some nonpolar analytes. Ionic analytes can be separated using ion-suppression or ion-pairing techniques, which will be discussed in Sections 2.3.4–2.3.6 in Chapter 2.

1.2.3 Ion-Exchange Chromatography (IEC)

In ion-exchange chromatography,[3–5] the separation mode is based on the exchange of ionic analytes with the counter-ions of the ionic groups attached to the solid support (Figure 1.6a). Typical stationary phases are cationic exchange (sulfonate) or anionic exchange (quaternary ammonium) groups bonded to polymeric or silica materials. Mobile phases consist of buffers, often with increasing ionic strength, to force the migration of the analytes. Common applications are the analysis of ions and biological components such as amino acids, proteins/peptides, and polynucleotides. Figure 1.7 shows the separation of amino acids on a sulfonated polymer column and a mobile phase of

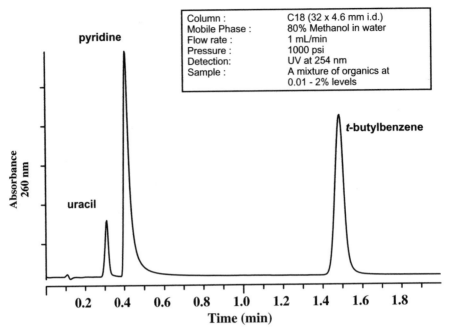

Figure 1.5. A reversed-phase HPLC chromatogram of three organic components eluting in the order of "polar first and nonpolar last." The basic pyridine peak is tailing due to a secondary interaction of the nitrogen lone-pair with residual silanol groups of the silica based bonded phase. Figure reprinted with permission from reference 8, Chapter 2.

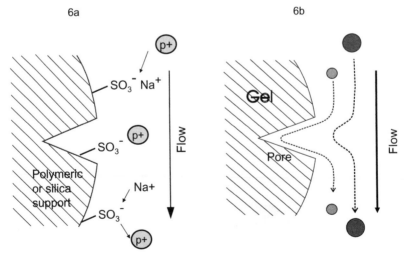

Figure 1.6. a. Schematic diagrams depicting separation modes of (a) ion-exchange chromatography (IEC), showing the exchange of analyte ion p^+ with the sodium counter ions of the bonded sulfonate groups; (b) size-exclusion chromatography (SEC), showing the faster migration of large molecules.

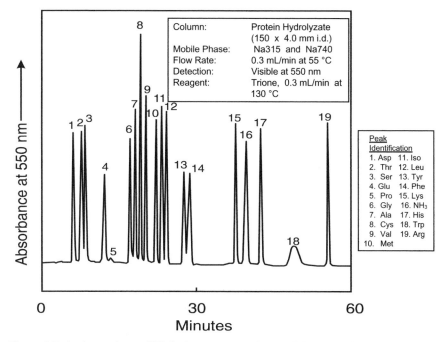

Figure 1.7. An ion-exchange HPLC chromatogram of essential amino acids using a cationic sulfonate column and detection with post-column reaction. Note that Na315 and Na740 are prepackaged eluents containing sodium ion and buffered at pH of 3.15 and 7.40, respectively. Trione is a derivatization reagent similar to ninhydrin. Chromatogram courtesy of Pickering Laboratories.

increasing sodium ion concentration and increasing pH. Since amino acids do not absorb strongly in the UV or visible region, a post-column reaction technique is used to form a color derivative to enhance detection at 550 nm. Ion chromatography[14] is a segment of IEC pertaining to the analysis of low concentrations of cations or anions using a high-performance ion-exhange column, often with a specialized conductivity detector.

1.2.4   Size-Exclusion Chromatography (SEC)

Size-exclusion chromatography[15] is a separation mode based solely on the analyte's molecular size. Figure 1.6b shows that a large molecule is excluded from the pores and migrates quickly, whereas a small molecule can penetrate the pores and migrates more slowly down the column. It is often called gel-permeation chromatography (GPC) when used for the determination of molecular weights of organic polymers and gel-filtration chromatography (GFC) when used in the separation of water-soluble biological materials. In GPC, the column is packed with cross-linked polystyrene beads of controlled pore sizes and eluted with common mobile phases such as toluene and tetrahy-

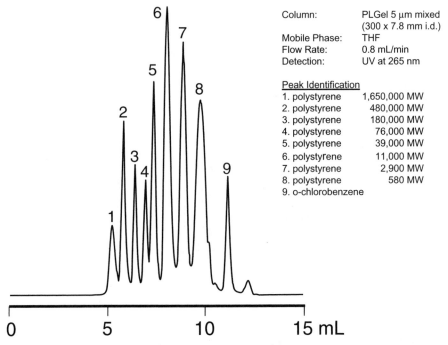

Column:	PLGel 5 µm mixed (300 x 7.8 mm i.d.)
Mobile Phase:	THF
Flow Rate:	0.8 mL/min
Detection:	UV at 265 nm

Peak Identification

1. polystyrene	1,650,000 MW
2. polystyrene	480,000 MW
3. polystyrene	180,000 MW
4. polystyrene	76,000 MW
5. polystyrene	39,000 MW
6. polystyrene	11,000 MW
7. polystyrene	2,900 MW
8. polystyrene	580 MW
9. o-chlorobenzene	

Figure 1.8. A GPC chromatogram of polystyrene standards on a mixed-bed polystyrene column. Chromatogram courtesy of Polymer Laboratories.

drofuran. Figure 1.8 shows the separation of polystyrene standards showing an elution order of decreasing molecular size. Detection with a refractive index detector is typical.

1.2.5 Other Separation Modes

Besides the four major HPLC separation modes, several others often encountered in HPLC or related techniques are noted below.

- *Affinity chromatography[9]:* Based on a receptor/ligand interaction in which immobilized ligands (enzymes, antigens, or hormones) on solid supports are used to isolate selected components from a mixture. The retained components can later be released in a purified state.
- *Chiral chromatography[16]:* For the separation of enantiomers using a chiral-specific stationary phase. Both NPC and RPC chiral columns are available.
- *Hydrophilic interaction chromatography (HILIC)[9]:* This is somewhat similar to normal phase chromatography using a polar stationary phase such as silica or ion-exchange materials but eluted with polar mobile

phases of organic solvents and aqueous buffers. It is most commonly used to separate polar analytes and hydrophilic peptides.

- *Hydrophobic interaction chromatography*[4,9]: Analogous to RPC except that mobile phases of low organic solvent content and high salt concentrations are used for the separation of proteins that are easily denatured by mobile phases with high concentrations of organic solvents used in RPC.

- *Electrochromatography:* Uses capillary electrophoresis[17] (CE) equipment with a packed capillary HPLC column. The mobile phase is driven by the electromotive force from a high-voltage source as opposed to a mechanical pump. It is capable of very high efficiency.

- *Supercritical fluid chromatography (SFC)*[18]: Uses HPLC packed columns and a mobile phase of pressurized supercritical fluids (i.e., carbon dioxide modified with a polar organic solvent). Useful for nonpolar analytes and preparative applications where purified materials can be recovered easily by evaporating the carbon dioxide. HPLC pumps and GC-type detectors are often used.

- *Other forms of low-pressure liquid chromatography:*
  — Thin-layer chromatography (TLC)[19] uses glass plates coated with adsorbents and capillary action as the driving force. Useful for sample screening and semi-quantitative analysis.
  — Paper chromatography (PC), a form of partition chromatography using paper as the stationary phase and capillary action as the driving force.
  — Flash chromatography, a technique for sample purification using disposable glass NPC columns and mobile phase driven by gas-pressure or low-pressure pumps.

1.3 SOME COMMON-SENSE COROLLARIES

The goal of most HPLC analysis is to separate analyte(s) from other components in the sample for accurate quantitation. Several corollaries are often overlooked by practitioners:

1. *Sample must be soluble:* "If it's not in solution, it cannot be analyzed by HPLC." Solubility issues often complicate assays of low-solubility analytes or component difficult to extract from sample matrices. Low recoveries often stem from poor sample preparation steps rather than the HPLC analysis itself.

2. *For separation to occur, analytes must be retained and have differential migration in the column:* Separation cannot occur without retention and sufficient interaction with the stationary phase. For quantitative analy-

sis, analytes must have different retention on the column versus other components.

3. *The mobile phase controls the separation:* Whereas the stationary phase provides a media for analyte interaction, the mobile phase controls the overall separation. In HPLC method development, efforts focus on finding a set of mobile phase conditions for separating the analyte(s) from other components. Exceptions to this rule are size exclusion, chiral, and affinity chromatography.

4. *All C18-bonded phase columns are not created equal and are not interchangeable:* There are hundreds of C18 columns on the market. They vary tremendously in their retention and silanol characteristics.[9]

5. *The final analyte solution should be prepared in the mobile phase:* The final analyte solution, if possible, should be dissolved in the mobile phase or a solvent of "weaker" strength than the starting mobile phase. Many anomalies such as splitting peaks or fronting peaks are caused by injecting samples dissolved in solvents stronger than the mobile phase. Inject a smaller injection volume (2–5 µL) if a stronger solvent must be used to minimize these problems.

6. *Every analytical method has its own caveats, limitations, or pitfalls:* An experienced method development scientist should identify these potential pitfalls and focus on finding conditions to minimize these problems areas for more reliable analysis.

1.4   HOW TO GET MORE INFORMATION

Beginners are encouraged to obtain more information from the following sources:

- Training courses sponsored by training institutions,[20] manufacturers, or national meetings (American Chemical Society, Pittsburgh Conference, Eastern Analytical Symposium).
- Computer-based training programs.[21]
- Useful books[3,9] and websites[22–24] of universities and other government or compendia agencies, such as the U.S. Food and Drug Administration (FDA), U.S. Environmental Protection Agency (EPA), International Conference on Harmonization (ICH), United States Pharmacopoeia, (USP), Association of Official Analytical Chemist International (AOAC), and American Society of Testing and Materials (ASTM).
- Research and review articles published in journals such as the *Journal of Chromatography, Journal of Chromatographic Science, Journal of Liquid Chromatography, LC.GC Magazine, Analytical Chemistry*, and *American Laboratories.*

1.5 SUMMARY

This introductory chapter describes the scope of the book and gives a brief summary of the history, advantages, limitations, and some common-sense axioms of HPLC. Major separation modes are discussed and illustrated with examples. Information resources on HPLC are also listed.

1.6 REFERENCES

1. C. Poole, *The Essence of Chromatography,* Elsevier, Amsterdam, 2003. (A 927-page book that provides a comprehensive survey of the current practice of chromatography. It includes an in-depth and well-referenced treatment of chromatographic theory and quantitative principles. It covers HPLC, GC, thin-layer chromatography, supercritical fluid chromatography, and capillary electrophoresis (CE).)

2. J. Miller, *Chromatography: Concepts and Contrasts*, 2nd Edition, Wiley, New York, 2004. (An updated text on all phases of chromatography, including HPLC, GC, CE, sampling, and sample preparation, as well as their industrial practices in regulated industries.)

3. L.R. Snyder and J.J. Kirkland, *Introduction to Modern Liquid Chromatography*, 2nd Edition, John Wiley & Sons, New York, 1979. (This is the second edition of the classic book published in 1973 on HPLC fundamentals and applications.)

4. L.R. Snyder, J.J. Kirkland, and J.L. Glajch, *Practical HPLC Method Development,* 2nd Edition, Wiley-Interscience, New York, 1997. (A comprehensive text on all phases on HPLC method development.)

5. V.R. Meyer, *Practical HPLC, 4th edition*, Wiley, New York, 2004. (This popular updated text provides a systematic treatment of HPLC and has broad appeal to students and laboratory professionals.)

6. E. Katz, R. Eksteen, P. Schoenmakers, and N. Miller, *Handbook of HPLC*, Marcel Dekker, New York, 1998. (This 1,000-page, comprehensive handbook covers fundamentals, instrumentation, and applications at great length and in theoretical depth.)

7. B. Bidlingmeyer, *Practical HPLC Methodologies and Applications,* Wiley Interscience, New York, 1993. (This book focuses on HPLC method development and applications.)

8. S. Ahuja, M.W. Dong, eds., *Handbook of Pharmaceutical Analysis by HPLC*, Elsevier, Amsterdam, 2005. (A reference guide on the practice of HPLC in pharmaceutical analysis.)

9. U.D. Neue, *HPLC Columns: Theory, Technology, and Practice*, Wiley-VCH, New York, 1997. (This book focuses on all phases of column technologies, including theory, column design, packing, chemistry, modes, method development, and maintenance.)

10. L.S. Ettre, *Milestone in the Evolution of Chromatography*, ChromSource, Portland, OR, 2002.

11. *HPLC: Opportunities in a Fragmented Market*, Strategic Directions International, Inc., Los Angeles, 2003.

12. W.R. Melander and Cs. Horvath, in Cs. Horvath, ed., *High-Performance Liquid Chromatography*: *Advances and Perspectives*, Volume 2, Academic Press, 1980, p. 113.

13. P.W. Carr, D.E. Martire, and L.R. Snyder, *J. Chromatogr. A*, **656**, 1 (1993).

14. J. Weiss, *Ion Chromatography*, 2nd Edition, VCH, Weinheim, 1995.

15. S. Mori, H.G. Barth, *Size Exclusion Chromatography*, Springer-Verlag, New York, 1999.

16. T.E. Beesley and R.P.W. Scott, *Chiral Chromatography*, John Wiley & Sons, New York, 1999.

17. R. Weinberger, *Practical Capillary Electrophoresis*, 2nd Edition, Academic Press, New York, 2000.

18. K. Anton, C. Berger, *Supercritical Fluid Chromatography with Packed Columns*: *Techniques and Applications*, Marcel Dekker, New York, 1997.

19. J.C. Touchstone, *Practice of Thin-Layer Chromatography*, 3rd Edition, Wiley-Interscience, New York, 1992.

20. *LC Resources, training courses*, Walnut Creek, CA: http://www.lcresources.com/training/trprac.html

21. *Introduction to HPLC*, *CLC-10* (Computer-based Instruction), Academy Savant, Fullerton, CA: http://www.savant4training.com/savant2.htm

22. http://hplc.chem.shu.edu (popular website by Prof. Y. Kazakevich of Seton Hall University on basic HPLC).

23. http://www.forumsci.co.il/HPLC/topics.html (by Dr. Shulamit Levin of Hebrew University with many useful links on HPLC and LC/MS).

24. http://www.separationsnow.com (a free-access web portal on all separation sciences, sponsored by John Wiley & Sons).

Note: Although the above web addresses are current at press time, they might change with time. The reader should use a search engine such as Google™ to locate new addresses or other useful websites.

2

BASIC TERMS AND CONCEPTS

Modern HPLC for Practicing Scientists, by Michael W. Dong
Copyright © 2006 John Wiley & Sons, Inc.

2.1 SCOPE

The objective of this chapter is to provide the reader with a concise overview of HPLC terminology and concepts. Both basic and selected advanced concepts are covered. The reader is referred to other HPLC textbooks,[1-7] training courses,[8-9] journals, and Internet resources for a more detailed treatment of HPLC theory and concepts. This chapter has the following sections:

- Basic terminology and concepts of retention, selectivity, efficiency, resolution, and peak tailing
- Mobile phase basics and parameters (solvent strength, pH, ion pairing reagent, flow, and temperature)
- The Resolution Equation (effect of efficiency, retention and selectivity)
- The van Deemter Equation (effect of particle size and flow rate)
- Concepts in gradient analysis (peak capacity, effects of flow rate, gradient time) and method orthogonality

The focus is on concepts in reversed-phase liquid chromatography (RPLC), though the same concepts are usually applicable to other modes of HPLC. International Union of Pure and Applied Chemistry (IUPAC)[10] nomenclature is used. The term "sample component" is often used interchangeably with "analyte" and "solute" in the context of this book. As mentioned in Chapter 1, the most common stationary phase is a hydrophobic C18-bonded phase on a silica support used with a mixed organic and aqueous mobile phase. The terms "packing" and "sorbent" often refer to the bonded phase whereas solid support refers to the unbonded silica material.

2.2   BASIC TERMS AND CONCEPTS

2.2.1   Retention Time (t_R), Void Time (t_M), Peak Height (h), and Peak Width (w_b)

Figure 2.1 shows a chromatogram with a single sample component. The time between the sample injection and the peak maximum is called the retention time (t_R). The retention time of an unretained component or the first baseline disturbance by the sample solvent is called the void time (t_M) or hold-up time. t_M is the total time spent by any component in the mobile phase. The adjusted retention time, t_R' is equal to ($t_R - t_M$), i.e., the time the solute resides in the stationary phase. Thus, $t_R = t_R' + t_M$ or the retention time is the total time the solute spends in the stationary phase (t_R') and in the mobile phase (t_M).

The solute peak has both a peak width and a peak height (h). The peak width is usually measured at the base (w_b) or at the peak half-height ($w_{1/2}$). Figure 2.2 shows how w_b and $w_{1/2}$ are measured. Two tangent lines are drawn from the steepest inflection points of the peak. The distance between the two points at which the two tangents intercept with the baseline is w_b. Note that peak area is roughly equal to $\frac{1}{2}(w_b \times h)$.[1,6] For Gaussian peaks, w_b is approximately equal to four times the standard deviation (4σ), which brackets 95% of the total peak area. The width at half height ($w_{1/2}$) is easier to measure and is usually used to calculate column efficiency.

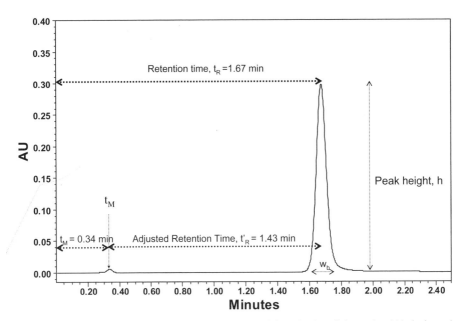

Figure 2.1. A chromatogram showing retention time (t_R), void time (t_M), peak width (w_b), and peak height (h).

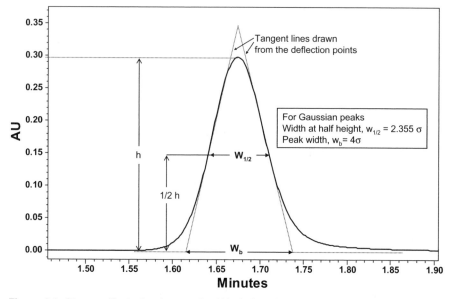

Figure 2.2. Diagram illustrating how peak width (w_b) and peak width at half height ($w_{1/2}$) are measured.

The height or the area of a peak is proportional to the amount of analyte component. The peak area is commonly used to perform quantitative calculations.

2.2.2   Retention Volume (V_R), Void Volume (V_M), and Peak Volume

The retention volume (V_R) is the volume of mobile phase needed to elute the analyte at given flow rate (F).
    Here,

$$\text{retention volume, } V_R = t_R F. \qquad \text{Eq. 2.1}$$

Similarly,

$$\text{void volume, } V_M = t_M F. \qquad \text{Eq. 2.2}$$

The void volume (V_M) is the total volume of the liquid mobile phase contained in the column (also called hold-up volume). It is the volume of the empty column (V_c) minus the volume of the solid packing. Note that V_M is the sum of the intraparticle volume (V_0) and the interstitial volumes (V_e) inside the pores of the solid support. For most columns, the void volume can be estimated by the equation

$$V_M = 0.65 \ V_c = 0.65 \ \pi \ r^2 \ L, \qquad\qquad \text{Eq. 2.3}$$

where r is the inner radius of the column and L is the length of the column. V_M can also be estimated from t_M in the chromatogram, since $V_M = t_M F$ (Eq. 2.2). Note that V_M is proportional to r^2, which dictates the operating flow rate through the column. Note also that V_o does not include the interstitial pore volume and is equal to V_M only for columns packed with nonporous particles.

The peak volume, also called bandwidth, is the volume of mobile phase containing the eluted peak:

$$\text{Peak volume} = w_b F. \qquad\qquad \text{Eq. 2.4}$$

Peak volume is proportional to V_M, and therefore smaller columns produce smaller peak volumes (See Section 2.2.6 and Eq. 2.14).

2.2.3 Retention Factor (k)

The retention factor (k) is the degree of retention of the sample component in the column. k is defined as the time the solute resides in the stationary phase (t_R') relative to the time it resides in the mobile phase (t_M), as shown in Figure 2.3. k, an IUPAC term, was often referred to as k' or capacity factor in many references.

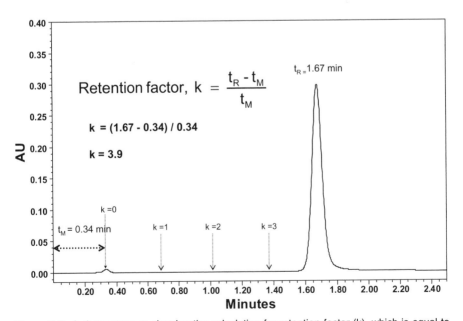

Figure 2.3. A chromatogram showing the calculation for retention factor (k), which is equal to t_R'/t_0. k is an important parameter defining the retention of the analyte. Desirable k values for isocratic analyses are 1 to 20.

$$\text{Retention factor, k} = \frac{t'_R}{t_M} = \frac{t_R - t_M}{t_M} \qquad \text{Eq. 2.5}$$

Rearranging this equation, we get

$$t_R = t_M + t_M k = t_M(1+k). \qquad \text{Eq. 2.6}$$

Equation 2.6 indicates that retention time is proportional to k.

Note that by multiplying both sides by the flow rate, F, a similar equation for V_R is obtained:

$$\text{Retention volume, } V_R = F \, t_R = F \, t_M(1+k) \quad or \quad V_R = V_M(1+k) \qquad \text{Eq. 2.7}$$

A peak with k = 0 is a component that is unretained by the stationary phase and elutes with the solvent front. k > 20 indicates that the component is highly retained. In most analyses, analytes elute with k between 1 and 20 so that they have sufficient opportunity to interact with the stationary phase resulting in differential migration. Analytes eluting with k > 20 are difficult to detect due to excessive band broadening. Figure 2.3 shows an example of how k is calculated from t_R and t_M.

Chromatography is a thermodynamically based method of separation, where each component in the sample is distributed between the mobile phase and the stationary phase.[7,11]

$$X_m \leftrightarrow X_s \quad \text{Partition coefficient, K} = \frac{[X_s]}{[X_m]} \qquad \text{Eq. 2.8}$$

where $[X_m]$ and $[X_s]$ are the concentrations of analyte X in the mobile phase and stationary phase, respectively. The distribution of analyte X is governed by the partition coefficient, K. k can also be described by the ratio of total number of moles of analytes in each phase[7]

$$\text{Retention factor, k} = \frac{Moles\ of\ X\ in\ stationary\ phase}{Moles\ of\ X\ in\ mobile\ phase}$$

$$= \frac{[X_s]}{[X_m]} \frac{V_s}{V_M} = K \frac{V_s}{V_M} \qquad \text{Eq. 2.9}$$

where V_s is the volume of the stationary phase and V_m is the volume of the mobile phase in the column or the void volume. k is primarily controlled by the strength of the mobile phase, the nature of stationary phase, and the temperature at which the separation is performed.

2.2.4 Separation Factor (α)

The separation factor or selectivity (α)[1,11] is a measure of relative retention k_2/k_1 of two sample components as shown in Figure 2.4. Selectivity must be

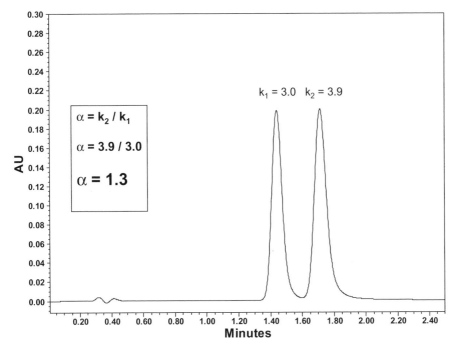

Figure 2.4. A chromatogram of two peaks with a selectivity factor (α) of 1.3.

>1.0 for peak separation. Selectivity is dependent on many factors that affect K such as the nature of the stationary phase, the mobile phase composition, and properties of the solutes. Experienced chromatographers can skillfully exploit the selectivity effects during method development to increase the separation of key analytes in the sample.[2]

2.2.5   Column Efficiency and Plate Number (N)

An efficient column produces sharp peaks and can separate many sample components in a relatively short time. As seen in most chromatograms, peaks tend to be Gaussian in shape and broaden with time, where w_b becomes wider with longer t_R. This band broadening inside the column is fundamental to all chromatographic processes.[1,6,12] The number of theoretical plates or plate number (N) is a measure of the efficiency of the column. N is defined as the square of the ratio of the retention time divided by the standard deviation of the peak (σ). Since w_b is equal to 4σ for a Gaussian peak,

$$\text{Number of theoretical plates, N} = \left(\frac{t_R}{\sigma}\right)^2 = \left(\frac{4t_R}{w_b}\right)^2 = 16\left(\frac{t_R}{w_b}\right)^2. \qquad \text{Eq. 2.10}$$

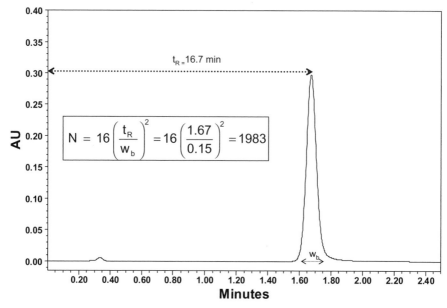

Figure 2.5. A chromatogram showing a peak from a column with N = 1,983.

Figure 2.5 shows an example of how N is calculated using the equation above. Since it is more difficult to measure σ or w_b, a relationship using width at half height ($w_{1/2}$) is often used to calculate N as described in the United States Pharmacopoeia (USP). Note that for a Gaussian peak, $w_{1/2}$ is equal to 2.355σ (Figure 2.2)[6]:

$$N = \left(\frac{t_R}{\sigma}\right)^2 = \left(\frac{2.355\, t_R}{w_{1/2}}\right)^2 = 5.546\left(\frac{t_R}{w_{1/2}}\right)^2. \qquad \text{Eq. 2.11}$$

2.2.6 Peak Volume

Peak volume is the volume of mobile phase or eluate containing the eluting peak. Peak volume is proportional to k and V_M. This relationship between peak volume and these factors can be derived by rearranging Eq. 2.10:

$$N = 16\left(\frac{t_R}{w_b}\right)^2 \qquad \frac{N}{16} = \left(\frac{t_R}{w_b}\right)^2 \qquad \frac{\sqrt{N}}{4} = \left(\frac{t_R}{w_b}\right), \qquad \text{Eq. 2.12}$$

then multiplying both numerator and denominator by F,

$$\frac{\sqrt{N}}{4} = \left(\frac{F\, t_R}{F\, w_b}\right) = \left(\frac{V_R}{F\, w_b}\right) = \left(\frac{V_R}{Peak\ Volume}\right). \qquad \text{Eq. 2.13}$$

Thus,

$$Peak\ volume = \frac{4V_R}{\sqrt{N}} = \frac{4\ V_M(1+k)}{\sqrt{N}}.$$ Eq. 2.14

Since for a given column and a set of operating conditions N is approximately constant and $(1+k)$ is roughly equal to k in most cases where k is much greater than 1, peak volume is proportional to k and V_M. This relationship is important because of the increasing use of smaller-diameter columns (column i.d. <3 mm) with smaller V_M, since the smaller peak volumes from these columns are highly affected by the deleterious effect of dispersion by the instrument (extra-column bandbroadening). This effect will be discussed section 4.10 of Chapter 4.

2.2.7 Height Equivalent to a Theoretical Plate or Plate Height (HETP or H)

The concept of a "plate" was adapted from the industrial distillation process using a distillation column consisting of individual plates where the condensing liquid is equilibrating with the rising vapor. A longer distillation column would have more "plates" or separation power and could separate a raw material such as crude oil into more fractions of distillates. Although there are no discreet plates inside the HPLC column, the same concept of plate number (N) or plate height (H) can be applied. The height equivalent to a theoretical plate (HETP or H) is equal to the length of the column (L) divided by the plate number (N):[6,7,12]

$$HETP, H = L/N.$$ Eq. 2.15

In HPLC, the main factor controlling H is the particle diameter of the packing (d_p). For a well-packed column, H is roughly equal to $2.5\ d_p$.

A typical 15-cm-long column packed with 5-μm materials should have $N = L/H = 150,000\,\mu m/(2.5 \times 5\,\mu m)$, or about 12,000 plates. Similarly, a typical 15-cm column packed with 3-μm material should have $N = L/H = 150,000\,\mu m/7.5\,\mu m$, or about 20,000 plates. Thus, columns packed with smaller particles are usually more efficient and have a higher plate number.

2.2.8 Resolution (R_s)

The goal of most HPLC analyses is the separation of one or more analytes in the sample from all other components present. Resolution (R_s) is a measure of the degree of separation of two adjacent analytes. R_s is defined as the difference in retention times of the two peaks divided by the average peak width (Figure 2.6). Since peak widths of adjacent peaks tend to be similar, the average peak width is approximated by one of the w_b's:

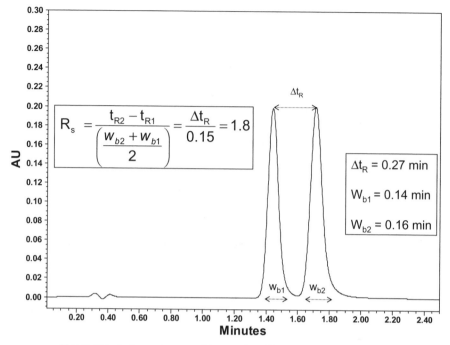

Figure 2.6. A chromatogram of two peaks with a resolution (R_s) of 1.8.

$$\text{Re solution, } R_s = \frac{t_{R2} - t_{R1}}{\left(\dfrac{w_{b1} + w_{b2}}{2}\right)} = \frac{\Delta t_R}{w_b}. \qquad \text{Eq. 2.16}$$

Figure 2.7 is a graphic representation of resolution for two peaks with R_s ranging from 0.6 to 2.0. Note that $R_s = 0$ indicates complete co-elution or no separation. $R_s = 0.6$ indicates that a shoulder is discernible or a slight partial separation. $R_s = 1$ indicates that a partial separation and is the minimum separation required for quantitation. $R_s = 1.5$ indicates baseline separation. Ideally, the goal of most HPLC methods is to achieve baseline separation ($R_s = 1.5$–2.0) for all key analytes.[2,3]

2.2.9 Peak Symmetry: Asymmetry Factor (A_s) and Tailing Factor (T_f)

Under ideal conditions, chromatographic peaks should have Gaussian peak shapes with perfect symmetry. In reality, most peaks are not perfectly symmetrical and can be either fronting or tailing (Figure 2.8). The asymmetry factor (A_s) is used to measure the degree of peak symmetry and is defined at peak width of 10% of peak height ($W_{0.1}$). Note that T_f is used here instead of T, as in the USP, because T often stands for temperature.

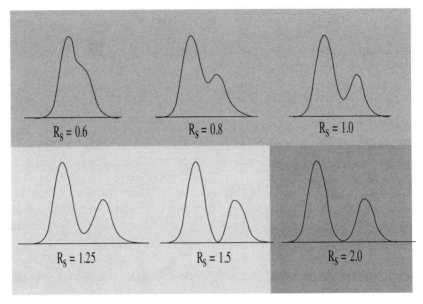

Figure 2.7. Diagrams showing two closely eluting peaks at various resolution values from 0.6 to 2.0. Figure reprinted with permission from Academy Savant.

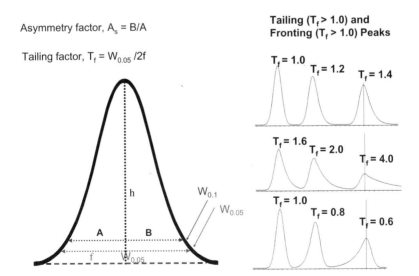

Figure 2.8. A diagram showing the calculation of peak asymmetry (A_s) and tailing factor (T_f) from peak width at 5% height ($W_{0.05}$) according to the USP. Inset diagrams show fronting and tailing peaks.

$$\text{Asymmetry factor, } A_s = B/A \quad \text{(see Figure 2.8).} \qquad \text{Eq. 2.17}$$

Tailing factor (T_f) is a similar term defined by the USP.[13] T_f is calculated using the peak width at 5% peak height $(W_{0.05})$:

$$\text{Tailing factor } T_f = W_{0.05}/2f \quad \text{(see Figure 2.8).} \qquad \text{Eq. 2.18}$$

Tailing factors are a required calculation in most pharmaceutical methods. $T_f = 1.0$ indicates a perfectly symmetrical peak. $T_f > 2$ indicates a tailing peak that is typically not acceptable due to difficulty in integrating the peak area precisely. For most peaks $(0.5 < T_f < 2.0)$, the values of A_s and T_f are fairly similar. For severely tailing peaks, A_s tends to be somewhat larger than T_f.

Peak tailing is typically caused by adsorption or extracolumn band broadening. Peak fronting is typically caused by column overloading or chemical reaction of the analyte during chromatography.[1] For instance, many basic analytes (amines) display some peak tailing due to the polar interaction with residual acidic silanol groups in silica-based columns.[6] Figure 2.9 shows an RPLC chromatogram with three components. Uracil is very soluble in the

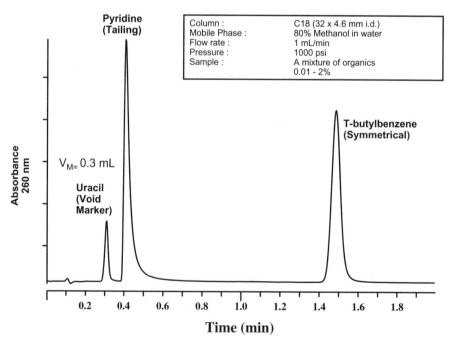

Figure 2.9. A HPLC chromatogram of three components with LC conditions shown in the inset. Note that the basic pyridine peak is tailing although the *t*-butylbenzene peak (neutral) is symmetrical. Reprint with permission from reference 3.

mobile phase and elutes with the solvent front with k = 0. Uracil is often used as a void volume marker for the measurement of V_M in RPLC. Pyridine is a base and exhibits considerable peak tailing due to secondary hydrophilic interaction with residual silanol groups in addition to the primary partitioning process with the C-18 bonded phase. *t*-Butylbenzene is a neutral and hydrophobic molecule, which elutes much later but with excellent peak symmetry.

2.3 MOBILE PHASE

The mobile phase is the solvent that moves the solute (analyte) through the column. In HPLC, the mobile phase interacts with both the solute and the stationary phase and has a powerful influence on solute retention and separation.[1–3,7]

2.3.1 General Requirements

Ideally, solvents used as HPLC mobile phases should have these characteristics:

- High solubility for the sample components
- Noncorrosive to HPLC system components
- High purity, low cost, UV transparency
- Other desirable characteristics include low viscosity, low toxicity, and non-flammability. Table 2.1 lists several common HPLC solvents and their important attributes.

Table 2.1. Common HPLC Solvents and Their Properties

Solvent	Solvent strength $(E°)$	bp (°C)	Viscosity (cP) at 20°C	UV cut-off (nm)	Refractive index
n-Hexane	0.01	69	0.31	190	1.37
Toluene	0.29	78	0.59	285	1.49
Methylene chloride	0.42	40	0.44	233	1.42
Tetrahydro-furan	0.45	66	0.55	212	1.41
Acetonitrile	0.55–0.65	82	0.37	190	1.34
2-Propanol	0.82	82	2.30	205	1.38
Methanol	0.95	65	0.54	205	1.33
Water	Large	100	1.00	<190	1.33

$E°$ (solvent elution strength as defined by Hildebrand on alumina). Data extracted from reference 2 and other sources.

2.3.2   Solvent Strength and Selectivity

Solvent strength refers to the ability of a solvent to elute solutes from a column.[1,2,7,11] Solvent strengths under normal phase conditions are often characterized by Hildebrand's elution strength scale (E°). Some are listed in Table 2.1. Solvent strength is related to its polarity. Nonpolar hexane is a weak solvent in normal phase chromatography whereas water is a strong solvent. The opposite is true in RPLC since the stationary phase is hydrophobic. Here water is a weak solvent and organic solvents are strong and in reversed order of the Hildebrand scale of THF > ACN > MeOH ≫ water. Water is a weak solvent because it is a poor solvent for nonstrongly H-bonding organics.

Figure 2.10 shows a series of six chromatograms to illustrate the effect of solvent strength in RPLC. Here, the two components (nitrobenzene and propylparaben) are eluted with mobile phases of decreasing solvent strength (i.e., decreasing concentration of acetonitrile (ACN)). At 100% ACN, both components are not retained by the column and elute with a k close to zero. At 60% ACN, the peaks are slightly retained (k close to 1) and are partially separated. The two components merge back together at 40% ACN. At 30% ACN, the two components are well separated, though propylparaben now

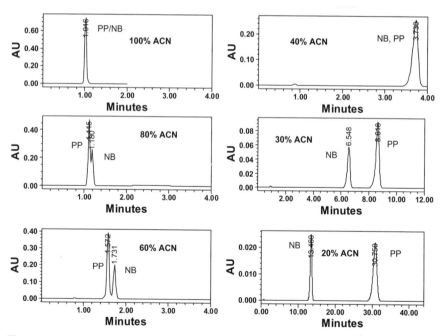

Figure 2.10. Six RPLC chromatograms illustrating the effect of mobile phase solvent strength on solute retention and resolution. LC conditions were: column: Waters Symmetry C18, 3 μm, 75 × 4.6 mm, 1 mL/min, 40°C, Detection at 258 nm. Mobile phase is mixture of acetonitrile (ACN) and water. Solutes were nitrobenzene (NB) and propylparaben (PP).

Table 2.2. Retention Data in Mobile Phases of Acetonitrile or Methanol for
Nitrobenzene (NB) and Propylparaben (PP)

% ACN	t_R (NB), min	k (NB)	t_R (PP), min	k (PP)	α (PP/NB)
100	1.02	0.28	1.02	0.28	1.00
90	1.04	0.30	1.04	0.30	1.00
80	1.18	0.48	1.12	0.39	0.83
70	1.38	0.73	1.27	0.59	0.81
60	1.73	1.16	1.57	0.96	0.83
50	2.37	1.96	2.29	1.86	0.95
40	3.73	3.66	3.73	3.66	1.00
30	6.55	7.19	8.62	9.78	1.36
25	9.25	10.56	15.35	18.19	1.72
20	13.46	15.83	30.75	37.44	2.37
% MeOH	t_R (NB), min	k (NB)	t_R (PP), min	k (PP)	α (PP/NB)
100	1.02	0.28	1.02	0.28	1.00
90	1.08	0.35	1.08	0.35	1.00
80	1.25	0.56	1.25	0.56	1.00
70	1.5	0.88	1.68	1.10	1.26
60	2.02	1.53	2.73	2.41	1.58
50	3.05	2.81	5.65	6.06	2.16
40	5.07	5.34	14.36	16.95	3.18
30	8.91	10.14	41	50.25	4.96
25	11.78	13.73	74	91.50	6.67

Column: Waters Symmetry C18, 3 μm, 75 × 4.6 mm, 1 mL/min, 40°C.
Solute: Nitrobenzene (NB), Propylparaben (PP).

elutes behind nitrobenzene. At 20% ACN, propylparaben is highly retained
with a k of 31.

Table 2.2 summarizes the t_R, k, and α of both nitrobenzene and propylparaben with percentage ACN and MeOH in the mobile phase. The following
observations can be made:

- Both t_R and k increase exponentially with decreasing percentage of
  organic solvents (or solvent strength) in the mobile phase.
- α and R_s generally increase with decreasing solvent strength.
- ACN is a stronger solvent than MeOH and can typically elute solutes
  faster in RPLC at similar concentration.

Figure 2.11 is a plot of the retention time and log k of nitrobenzene versus percentage ACN. Note that, as is typical in RPLC, log k is inversely proportional
to solvent strength or percentage of organic solvent. A useful rule of thumb
in RPLC indicates that a 10% decrease in the organic solvent in the mobile
phase typically produce affect a 3-fold increase in k or retention time.[2]

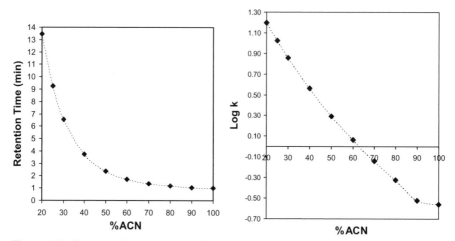

Figure 2.11. Retention data plots for nitrobenzene versus percentage acetonitrile of the mobile phase. Note that log k is inversely proportional to percentage acetonitrile. The proportionality is linear over a wide range. LC conditions were same as Figure 2.10.

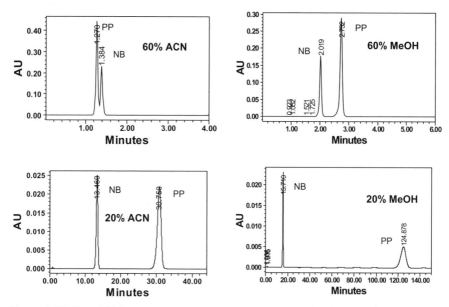

Figure 2.12. Four RPLC chromatograms illustrating the effect of mobile phase strength and selectivity of acetonitrile (ACN) and methanol (MeOH). See Figure 2.10 for LC conditions.

Figure 2.12 shows the chromatograms of the same two solutes eluted with 60% and 20% of ACN or MeOH, respectively. Notice the considerable differences of retention time and selectivity (elution order at 60% organic solvent) of the two solutes. Selectivity is a function of dipole moment, H-bonding, and dispersive characteristics of the stationary phase, the solute, and mobile phase components. Chapter 9 discusses how these selectivity effects can be manipulated in HPLC method development.

2.3.3 Buffers

The pH of the aqueous component in the mobile phase can have a dramatic effect on the retention of ionizable (acidic or basic) analytes. In RPLC, the ionized form of the solute does not partition well into the hydrophobic stationary phase and has significantly lower k than the neutral form. Figure 2.13 shows the retention map of two basic drugs. The k of the analyte is plotted against the pH of the mobile phase (with percentage organic solvent in the mobile phase unchanged). Note that at pH 2.0, both ionized solutes are not retained and elute as a single peak. At pH 8, the solutes are partially ionized and separate well. At pH 10, both un-ionized solutes are highly retained and resolved. Buffers are required to control the pH of the mobile phase. Table 2.3 summarizes the common buffers for HPLC and their respective pK_a and UV cutoffs. Buffers of ammonium salts of volatile acids are used for the development of mass spectrometer (MS) compatible HPLC methods. Note that buffers are only effective within ±1.5 pH unit from their pK_a. Although 50 mM buffers are specified in many older methods, the modern trend is to use low

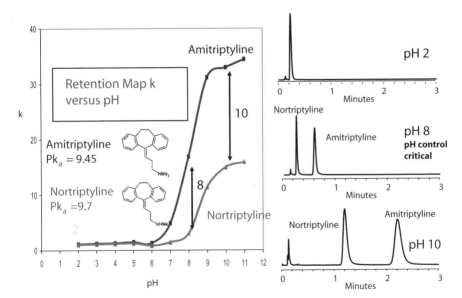

Figure 2.13. Retention map and chromatograms of two basic antidepressants using mobile phases at various pH with percentage organic modifier being kept constant. The diagram illustrates the importance of pH in the separation of basic analytes. Figure reprinted with permission from Waters Corporation.

Table 2.3. Common HPLC Buffers and Their Respective pK$_a$ and UV Cut-off

Buffer	pK$_a$	UV cut-off (nm)
Trifluoroacetic acid*	0.3	210
Phosphate	2.1, 7.2, 12.3	190
Citrate	3.1, 4.7, 5.4	225
Formate*	3.8	200
Acetate*	4.8	205
Carbonate*	6.4, 10.3	200
Tris(hydroxymethyl) aminomethane	8.3	210
Ammonia*	9.2	200
Borate	9.2	190
Diethylamine	10.5	235

*Volatile buffer systems, which are MS-compatible.
Data extracted from reference 2 and other sources.

buffer strengths, typically in the range of 10–20 mM.[12] Note that a disadvantage of using these volatile buffers is the loss of sensitivity at far UV (<230 nm) due to the inherent absorptivity of these buffers at these wavelengths.

2.3.4   Acidic Mobile Phases

In RPLC, acidic pH of 2.5–3 is used for many applications. The low pH suppresses the ionization of weakly acidic analytes, leading to higher retention.[2,3,11] Surface silanols are not ionized at low pH, lessening tailing with basic solutes. Most silica-based bonded phases are not stable below pH 2 due to acid catalyzed hydrolytic cleavage of the bonded groups.[2,3] Common acids used for mobile phase preparations are phosphoric acid, trifluoroacetic acid (TFA), formic acid, and acetic acid. However, basic analytes are ionized at low pH and might not be retained.

2.3.5   Ion-Pairing Additives

Ion-pairing reagents are detergent-like molecules added to the mobile phase to provide retention of acidic or basic analytes.[1,2,7] Long-chain alkyl sulfonates (C$_5$ to C$_{12}$) combine with basic solutes under acidic pH conditions to form neutral "ion-pairs" that are retained in RPLC. Retention is proportional to the length of the hydrophobic chain of the ion-pairing agent and its concentration. Note that TFA has some ion-pairing capability and is particularly useful in RPLC of proteins and peptides. Heptafluorobutyric acid (HFBA) is another useful volatile ion-pairing reagent that is also compatible with mass spectrometers. For acidic analytes, ion-pairing reagents such as tetraalkylammonium salts are used.

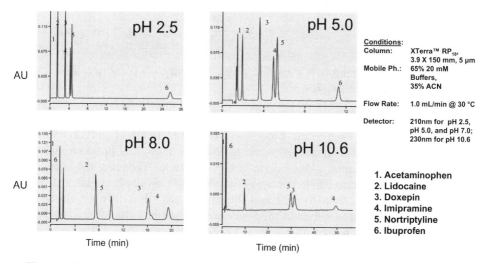

Figure 2.14. Selectivity effect of mobile phase pH. Chromatograms courtesy of Waters Corporation.

2.3.6   High pH Mobile Phases

Prior to 1990, the use of a high-pH mobile phase was not feasible with silica-base columns due to the dissolution of the silica support at pH > 8. The development of improved bonding chemistries and hybrid particles now extends the useful pH range from 2 to 10 or, in many cases, from 1 to 12 (see Chapter 3). This offers an important alternative approach for the separation of basic analytes and, in particular, for impurity testing of water-soluble basic drugs.[3,14] Figure 2.13 illustrates the basis of this approach in the separation of two closely related basic drugs, amitriptyline and nortriptyline. At low pH, both analytes are ionized and coelute with the solvent front. At pH close to the pK_a of the analytes, the partially ionized solutes are well separated with a large selectivity (α) value. At high pH, the non-ionized solutes are well retained and resolved. The advantages of high-pH separation versus ion-pairing are MS compatibility if a volatile buffer is used, better sensitivity, and excellent selectivity for closely related analytes[14] (see Section 8.8.3 in Chapter 8). Figure 2.14 illustrates the tremendous effects of mobile phase pH on solute retention and selectivity of a mixture of acidic and basic drugs.

2.3.7   Other Operating Parameters: Flow Rate (F) and Column Temperature (T)

Typical flow rates for analytical columns (4.6 mm i.d.) are 0.5–2 mL/min. Operating at higher flow rates (F) increases column back pressure (ΔP)[6] but reduces retention time and analysis time:

$$\Delta P = 1000 \frac{F\eta L}{\pi r^2 d_p^2} \qquad \text{Eq. 2.19}$$

where L = column length, η = mobile phase viscosity, r = column radius, and d_p = packing particle diameter.

For isocratic analysis where the mobile phase composition remains constant throughout the analysis, flow rate has no impact on k or α, since flow has the same effect on t_R of each solute. Flow also has significant effect on N, as shown in a Section 2.5 of this chapter. Operating flow rate should be proportional to the square of the column inner diameter. For instance, reducing the column diameter from 4 mm to 2 mm, the operating flow rate should be reduced from 1 mL/min to 0.25 mL/min to maintain the same linear flow rate, resulting in significant reduction of solvent usage for small-diameter columns.

Higher column temperatures (T) lower the viscosity of the mobile phase (thus, column back pressure, see Eq. 2.19) and usually have significant effects on retention (k), efficiency (N), and selectivity (α). Some of these effects are discussed further in Sections 2.4 and 2.5 of this chapter and in Chapter 8.

2.4   THE RESOLUTION EQUATION

The degree of separation or resolution (R_s) between two solutes is dependent on both thermodynamic factors (retention, k, and selectivity, α) and kinetic factors (peak width and column efficiency, N).[1-3,6-7,12] Resolution is controlled by three somewhat independent factors (retention, selectivity, and efficiency) as expressed quantitatively in the resolution equation:

$$\text{The Resolution Equation} \quad R_S = \left(\frac{k}{k+1}\right)\left(\frac{\alpha-1}{\alpha}\right)\left(\frac{\sqrt{N}}{4}\right) \qquad \text{Eq. 2.20}$$

To maximize R_s, k should be relatively large, though any k values >10 will drive the retention term of k / (k + 1) to approach unity. No separation is possible if k = 0, since R_s must equal zero if k is zero in the resolution equation. Selectivity (α) is typically between 1.01 and 1.50 for closely eluting solutes. When α = 1, R_s will equal to zero and, again, co-elution of the analytes occur. Selectivity is maximized by optimizing column and mobile phase conditions during method development. Figure 2.12 illustrates how resolution can be enhanced by exploiting the selectivity effect of the mobile phase (i.e., by switching from 60% ACN to 60% MeOH). Note that small change of selectivity can have a major effect on resolution as resolution is proportional to (α − 1). Columns of different bonded phases (i.e., C8, phenyl, CN, polar-embedded, see Chapter 4)[2,6] can also provide different selectivity effects. Finally, the plate number of the column (N) should be maximized by using a

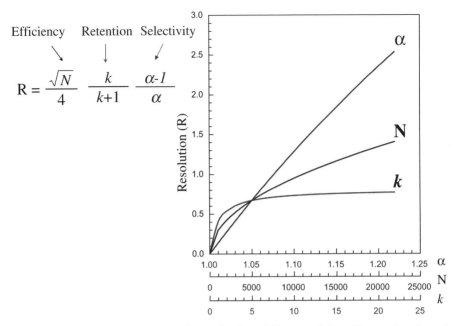

Figure 2.15. Graph illustrating the effects of α, k, and N on resolution. Diagram Courtesy of Supelco Inc.

longer column or a more efficient column. Note that increasing N is not an efficient way to achieve resolution since R_s is proportion to $\sqrt{N}$. Doubling N by doubling the column length increases analysis time by 2 but only increases resolution by $\sqrt{2}$ or by 41%. In contrast, increasing α from 1.05 to 1.10 will almost double the resolution. Nevertheless, for complex samples with many peaks, increasing N is a viable and the most direct approach. Figure 2.15 shows a chart summarizing the effect of α, k, and N on resolution. Figures 2.16 and 2.17 show examples on the effect of k and α on peak resolution. In Figure 2.16, seven substituted phenols in a test mixture can be separated by lowering the solvent strength from 45% to 30% ACN. Figure 2.17 illustrates how two analytes can be separated more effectively by changing the organic solvent in the mobile phase from acetonitrile to methanol or tetrahydrofuran through increasing the selectivity (α) of the separation.

2.5 THE VAN DEEMTER EQUATION

The van Deemter equation was developed in the 1950s to explain band broadening in chromatography by correlating HETP or plate height with linear flow velocity (V).[16] Figure 2.18 shows a typical van Deemter curve (HETP vs. V),

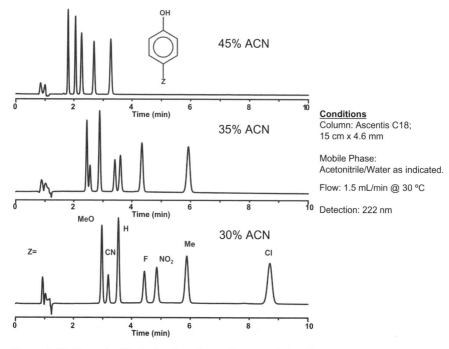

Figure 2.16. Examples illustrating the effect of k on resolution. Both retention (k) and resolution typically are increased by using lower solvent strength mobile phase as shown in the separation of seven substituted phenols. Diagram courtesy of Supelco Inc.

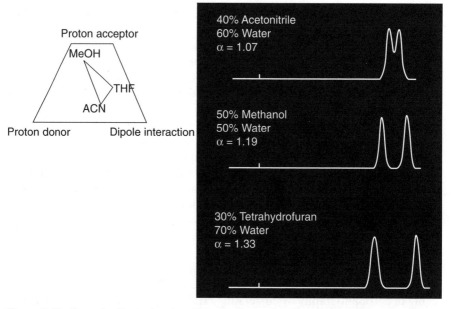

Figure 2.17. Examples illustrating the effect of solvent selectivity on resolution using different organic modifiers under RPLC conditions. The inset shows that the three typical reversed phase organic modifiers (ACN, MeOH, and THF) have very dissimilar solvent properties. Diagram courtesy of Academy Savant.

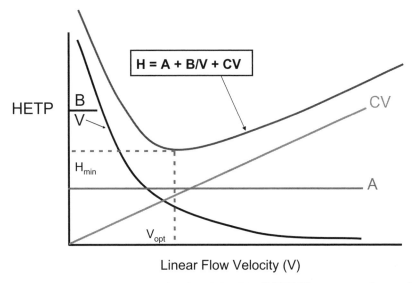

Figure 2.18. Van Deemter curve showing the relationship of HETP (H) vs. average linear velocity. The van Deemter curve has a classical shape and is a composite plot of A, B/V, and CV terms (plotted below to show their contributions). H_{min} = minimum plate height, V_{opt} = optimum velocity.

which is a composite curve from three relatively independent terms, which are in turn controlled by particle size (d_p) of the packing, and diffusion coefficients (D_m) of the solute.[1,2,5,6] The dip, or minimum point on the van Deemter curve, marks the minimum plate height (H_{min}) and the optimum velocity, which is the flow velocity at maximum column efficiency or H_{min}.

The van Deemter Equation $HETP = A + B/V + CV.$ Eq. 2.21

Figure 2.19 illustrates diagrammatically the causes of the various terms in the van Deemter equation.

- The A term represents "eddy diffusion or multi-path effect" experienced by the various solute molecules as they transverse the packed bed. A term is proportional to d_p and is smaller in well-packed columns.
- The B term represents "longitudinal diffusion" of the solute band in the mobile phase and is proportional to D_m of the solute. Note that contribution from the B term is only important at very low flow rate.
- The C term represents "resistance to mass transfer" due to time lags caused by the slower diffusion of the solute band in and out of the stationary phase. The C term is proportional to (d_p^2/D_m) and becomes

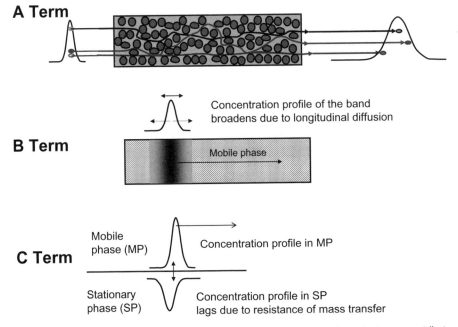

Figure 2.19. Diagrams illustrating the mechanism of the various van Deemter terms contributing to chromatographic band broadening. These diagrams are adapted from ideas from various Web resources.

important at high flow rates, especially for columns packed with larger particles.

Figure 2.20 shows the effect of d_p on van Deemter curves for columns packed with 10-, 5-, and 3-μm particles.[6,16,17] The following observations can be made.

- Small d_p yields smaller H_{min} (or a small particle column has more efficiency per unit length). Note that H is approximated by 2.5 d_p.
- Small-particle columns suffer less efficiency loss at high flow rates since the van Deemter curved at high flow is dominated by contribution from the C term, which is in turn proportional to d_p^2.

For these reasons, smaller-particle columns (i.e., 3 μm) are becoming popular in modern HPLC because of their inherent higher efficiency.[17] They are particularly useful in Fast LC and in high-speed applications such as high-throughput screening. In the last decade, sub-3-μm particles (i.e., 1.5–2.5 μm) were developed for even higher column performance, as discussed in Chapter 3.

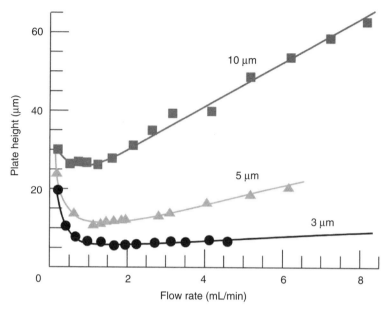

Figure 2.20. Experimental van Deemter curves of three columns packed with 10-, 5-, and 3-μm particles. Diagram reprinted with permission from reference 17.

2.6 ISOCRATIC VS. GRADIENT ANALYSIS

Most HPLC separations are performed under isocratic conditions in which the same mobile phase is used throughout the elution of the entire sample. Although isocratic analysis is good for simple mixtures, gradient analysis, in which the strength of the mobile phase is increased with time during sample elution, is preferred for more complex samples containing analytes of diverse polarities.[1-5]

Advantages of gradient analysis are:

- Better suited for complex samples and applications that require quantitation of all peaks or multiple analytes of diverse polarities
- Better resolution of early and late eluting peaks
- Better sensitivity of late eluting peaks
- Higher peak capacity (fit more peaks in the chromatogram)

Disadvantages are:

- More complex HPLC instrument (i.e., binary pump) is required
- Method development, implementation, and transfer are more difficult
- Typically longer assay times since column must be equilibrated with the initial mobile phase

There are several additional parameters in gradient analysis[2,3] not present in isocratic HPLC that need to be optimized. These are initial and final mobile phase composition, gradient time or duration (t_G), flow (F), and sometimes gradient curvature (linear, concave, and convex). Optimization of all these parameters is not intuitive but can often be readily accomplished by software simulation programs.[2]

2.6.1 Peak Capacity (n)

Unlike isocratic analysis where peaks are broadened with elution time, in gradient HPLC peaks have similar peak widths since they are eluted with an increasingly stronger mobile phase. Thus, column efficiency under gradient conditions cannot be measured using the usual equations 2.11 and 2.12 developed for isocratic separations. Peak capacity (n)[2,19] is the maximum number of peaks that can fit in a chromatogram with resolution value of one. Peak capacity is a useful concept for comparing column performance under different gradient conditions. Figure 2.21 shows that peak capacities can be approximated by t_G/w_b. Higher peak capacities of 100–200 are possible in gradient analysis versus typical peak capacities of 50–100 for isocratic analysis. This is a result of narrower peak widths typically achieved in gradient HPLC.

$$\text{Peak Capacity, } n = t_G/w_b \qquad \text{Eq. 2.22}$$

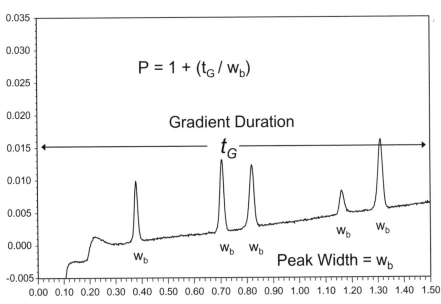

Figure 2.21. Chromatogram illustrating the concept of peak capacity (n), which is the maximum number of peaks that can be accommodated in a chromatogram with a resolution of one. Diagram courtesy of Waters Corporation.

2.6.2   Key Gradient Parameters (Initial and Final Solvent Strength, Gradient Time (t_G), and Flow Rate)

Gradient methods are more difficult to develop because the separation is controlled by several additional parameters, such as starting and ending solvent strength, flow rate (F), and gradient time (t_G). The concept of retention factor k is also more complex in gradient analysis and is best represented by an average k or k*,[2,19,20]

$$k* = \frac{t_G F}{1.15\, S\, \Delta\phi\, V_M},$$

where k* = average k under gradient conditions, $\Delta\phi$ = change in volume fraction of strong solvent in RPLC, S = a constant that is close to 5 for small molecules, F = flow rate (mL/min), t_G = gradient time (min), and V_M = column void volume (mL).

Unlike isocratic analysis, in gradient analysis, F has a dramatic influence on retention (k*). For instance, operating at higher F increases k*, since a greater volume of a lower-strength mobile phase is pumped through the column. This is equivalent to operating at a lower F and longer t_G. Figure 2.22 compares

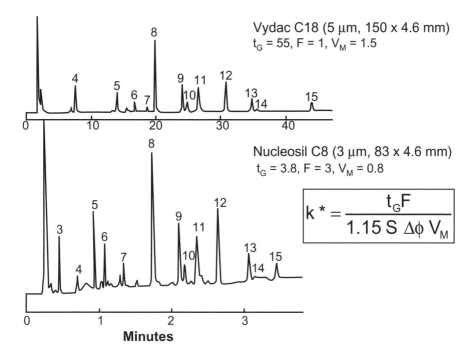

Figure 2.22. Two HPLC gradient chromatograms (tryptic maps of lysozyme) illustrating the dramatic effect of flow rate (F), gradient time (t_G), and holdup volume (V_M) on analysis time. Figure reprinted with permission from reference 20.

chromatograms for two gradient HPLC peptide maps of a protein digest[21] showing how a dramatically shorter analysis time can be achieved by reducing t_G, increasing F, and using a smaller column (lower V_M).

2.6.3   The 0.25 Δt_G Rule: When Is Isocratic Analysis More Appropriate?

Isocratic methods are typically used for the quantitation of a single analyte in the sample (or several analytes of similar polarities). Gradient methods are often required for multicomponent assays of complex samples or screening new samples of unknown composition. A rule of thumb[22] called the "0.25 Δt_G rule" is useful for checking whether a sample run under gradient conditions can be more effectively handled by isocratic analysis. In this approach, the sample is first analyzed under RPLC using a broad gradient (i.e., 5–95% organic solvent or MPB). If all analyte peaks elute within a time span of 0.25 Δt_G, then isocratic analysis is preferred. If the elution span is above 0.4 Δt_G, gradient is necessary. Between 0.25 Δt_G and 0.4 Δt_G, gradient is most likely preferred. Figure 2.23 shows an example illustrating this useful rule.

2.7   CONCEPT OF ORTHOGONALITY

How can one be assured that all the samples components are resolved? This is particularly important in critical assays such as pharmaceutical impurity testing to ensure that no impurity peaks are co-eluting with other components or hidden under the active ingredient peak. The standard practice is the use of an "orthogonal" separation technique or method to demonstrate that all impurities are accounted for. An orthogonal method is one based on different separation mechanism from the primary method.

Some examples of orthogonal methods are shown in Table 2.4. While the most extreme differences in separation mode and mechanism are found by using different techniques (HPLC vs. CE, GC, or SFC) or difference HPLC modes (RPLC vs. NPC, IEC, or HILIC), most practitioners prefer to use different variants of RPLC such as RPLC at different pH's, with or without ion pair reagent or using different columns (C18 vs. phenyl, cyano, or polar-embedded). This approach works better in practice under gradient conditions since the overall sample profile is preserved while substantial selectivity difference leads to changes of elution of order for most peaks. In many cases, mass spectrometric compatibility is also preserved. Figure 2.24 shows the column selectivity plots for a variety of different analytes. Since C8- and C18-bonded phases are very similar, the log k data of the two columns are well correlated for most analytes. Thus, methods using C8 and C18 columns are expected to yield similar elution profiles and are not orthogonal. In contrast, the selectivity differences of a C18 and a polar-embedded phase (amide) column lead to very scattered correlation. Methods using a C18 and a polar-

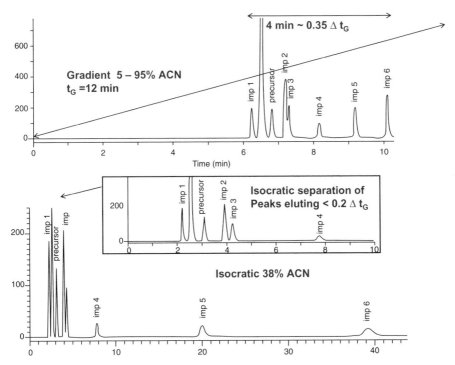

Figure 2.23. An example illustrating the 0.25 Δt_G rule used to indicate when a gradient separation can be more effectively carried out under isocratic conditions. The top chromatogram is obtained under gradient condition where all eight peaks elute in 4 minutes or 0.35 Δt_G. The bottom chromatogram shows the separation of these eight components under isocratic conditions. If all peaks elute <0.25 Δt_G in the gradient chromatogram, isocratic separation is preferred (as shown in the inset where the first six peaks are well-separated under isocratic conditions of 38% ACN. If all peaks eluting between 0.25 and 0.4 Δt_G, then both isocratic and gradient can be used. If all peaks elute in a time span >0.4 Δt_G, gradient separation is necessary.

Table 2.4. Common Examples of Orthogonal Separation Techniques

Category	Primary	Orthogonal
Technique	HPLC	CE, GC, SFC
HPLC Mode	RPLC	IEC, NPC, HILIC
Variants of RPLC	RPLC	RPLC with ion pairing
	RPLC at low pH	RPLC at high pH
	RPLC with C8 or C18 columns	RPLC with CN, phenyl, or polar-embedded columns

CE = capillary electrophoresis, SFC = supercritical fluid chromatography, HILIC = hydrophilic interaction chromatography.

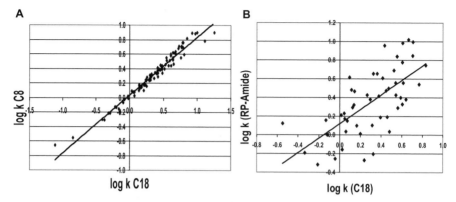

Figure 2.24. The concept of orthogonality as shown by retention plots of two sets of columns for a variety of different analytes. (A) Since the log k data of the two columns (C8 and C18) are well correlated for most analytes, these two columns are expected to yield similar elution profiles. (B) The selectivity differences of a C18 and a polar-embedded phase (amide) column lead to very scattered correlation of their respective retention data. Methods using a C18 and a polar-embedded column are therefore termed "orthogonal" and expected to yield very dissimilar profiles. Diagram courtesy of Supelco, Inc.

embedded column are "orthogonal" and expected to yield very dissimilar profiles. Another example can be found in Figure 8.8.

2.8 SAMPLE CAPACITY

Sample capacity or loading capacity of the column is the maximum amount of solute in milligrams per gram of packing material that can be injected without significantly reducing column efficiency[1,2] (e.g., by 10%). Since the amount of packing material is proportional to the column volume or size, sample capacity is proportional to the square of column diameter and the column length. It is therefore important in preparative work to maximize the yield of purified material in each sample run by increasing the column size.

2.9 GLOSSARY OF HPLC TERMS

Asymmetry (A_s)—Factor describing the shape of a chromatographic peak (Eq. 2.17).

Band Broadening—The process of increasing peak width and dilution of the solute as it travels through the column.

Column—A tube that contains packed solid material containing the stationary phase. Typical HPLC columns are stainless steel tubes packed with silica-based bonded phases.

Efficiency or Plate Number (N)—A measure of column performance. N is calculated from retention times and peak widths (Eq. 2.10 and 2.11).

Mobile Phase—A solvent that carries the sample through the column. Typical mobile phases in RPLC are mixtures of water with acetonitrile or methanol.

Plate Height (H)—Height equivalent to a theoretical plate as defined by dividing the column length by N. H_{min} is approximately equal to 2.5 d_p for a well-packed column (Eq. 2.15).

Peak Capacity (n)—The maximum number of peaks that can be resolved in a chromatogram with a resolution of one (Eq. 2.22).

Peak Width (w_b)—The width at the base of a chromatographic band during elution from the column. Higher-efficiency columns typically yield smaller-peak widths.

Resolution (R_s)—The degree of separation between two sample components as defined by the difference of their retention times divided by their average peak width (Eq. 2.16).

Retention—The tendency of a solute to be retained or retarded by the stationary phase in the column.

Retention Factor (k)—A measure of solute retention obtained by dividing the adjusted retention time by the void time. Also known as capacity factor or k' (Eq. 2.5).

Sample Capacity—The maximum mass of sample that can be loaded on a column without decreasing column efficiency.

Separation Factor or Selectivity (α)—The ratio of retention factors (k) of two adjacent peaks.

Stationary Phase—The immobile phase responsible for retaining the sample component in the column. In RPLC, this is typically the layer of hydrophobic groups bonded on silica solid support materials.

Void Volume (V_M)—The total volume of liquid held up in the column. It can be approximated as 65% of the volume of the empty column (Eq. 2.3).

A complete glossary of liquid-phase separation terms is found in reference 10.

2.10 SUMMARY AND CONCLUSION

This chapter provides an overview of basic terminology and essential concepts in HPLC including retention, selectivity, efficiency, resolution, and peak symmetry as well as their relationships with key column and mobile phase parameters. The resolution and van Deemter equations are discussed. The concepts of peak capacity and method orthogonality as well as key gradient parameters such as gradient time and flow rate are described. An abbreviated glossary of HPLC terms is listed.

2.11 REFERENCES

1. L.R. Snyder and J.J. Kirkland, *Introduction to Modern Liquid Chromatography*, John Wiley & Sons, New York, 1979.

2. L.R. Snyder, J.J. Kirkland, and J.L. Glajch, *Practical HPLC Method Development*, 2nd Edition, Wiley-Interscience, New York, 1997.

3. S. Ahuja and M.W. Dong, eds., *Handbook of Pharmaceutical Analysis by HPLC*, Elsevier/Academic Press, Amsterdam, 2005.

4. V.R. Meyer, *Practical HPLC*, 4th Edition, Wiley, New York, 2004.

5. E. Katz, R. Eksteen, P. Schoenmakers, and N. Miller, *Handbook of HPLC*, Marcel Dekker, New York, 1998.

6. U.D. Neue, *HPLC Columns: Theory, Technology, and Practice*, Wiley-VCH, New York, 1997.

7. T.H. Stout and J.G. Dorsey, in L. Ohannesian, and A.I. Streeter (eds.), *Handbook of Pharmaceutical Analysis*, Marcel Dekker, New York, 2001, p. 87.

8. Introduction to HPLC, CLC-10 and HPLC Method Development, CLC-20, (Computer-based Instruction), Academy Savant, Fullerton, CA: http://www.savant4training.com/savant2.htm

9. LC Resources, training courses, Walnut Creek, CA: http://www.lcresources.com/training/trprac.html

10. R.E. Majors and P.W. Carr, *LC.GC*, **19(2)**, 124 (2001).

11. A. Rizzi, in E. Katz, R. Eksteen, P. Schoenmakers, and N. Miller, eds., *Handbook of HPLC*, Marcel Dekker, New York, 1998.

12. R. Tijssen, in E. Katz, R. Eksteen, P. Schoenmakers, and N. Miller, eds., *Handbook of HPLC*, Marcel Dekker, New York, 1998.

13. *United States Pharmacopeia 26—National Formulatory 21*, The United States Pharmacopeial Convention, Rockville, MD, 2003.

14. M.W. Dong, G. Miller, and R. Paul, *J. Chromatog.* **987**, 283 (2003).

15. J.H. Zhao and P.W. Carr, *Anal. Chem.*, **71(14)**, 2623 (1999).

16. J.J. Van Deemter, F.J. Zuidenweg, and A. Klinkenberg, *Chem. Eng. Sci.* **5**, 271 (1956).

17. M.W. Dong and J.R. Gant, *LC.GC*, **2(4)**, 294 (1984).

18. M.W. Dong, *Today's Chemist at Work*, **9(2)**, 46 (2000).

19. U.D. Neue and J.R. Mazzeo, *J. Sep. Sci.*, **24**, 921 (2001).

20. L.R. Snyder and M. A. Stadalius, in Cs. Horvath, ed., *High-Performance Liquid Chromatography: Advances and Perspectives*, Volume 4, Academic Press, New York, 1986, p. 195.

21. M.W. Dong, in P. Brown, ed., *Advances in Chromatography*, Volume 32, Marcel Dekker, New York, 1992, p. 21.

22. Training course on "Advanced HPLC Method Development," LC Resources, Walnut Creek, CA. (www.lcresources.com)

3

HPLC COLUMNS AND TRENDS

Modern HPLC for Practicing Scientists, by Michael W. Dong
Copyright © 2006 John Wiley & Sons, Inc.

3.1 SCOPE

The HPLC column is the heart of the HPLC system. It holds the fine support media for the stationary phase that provides differential retention of sample components. This chapter offers a comprehensive but concise overview of HPLC columns and their modern trends. Basic HPLC column parameters such as column types (silica, polymer), modes, dimensions (preparative, analytical, narrowbore, micro LC), and packing characteristics (particle and pore size, bonding chemistry) are presented with key column development trends (high-purity silica, hybrid particles, novel bonding chemistries). The benefits of shorter and/or narrower columns packed with smaller particles (Fast LC, micro LC) are described. Guard columns and specialty columns are discussed together with column selection suggestions. For more detailed discussions on column technologies, the reader is referred to Uwe Neue's comprehensive book on HPLC columns,[1] other textbooks,[2–7] and journal articles.[8] Other useful resources include manufacturers' websites on their current column products. As in other chapters, the focus is on reversed-phase HPLC columns of small molecules, which represent 70–80% of all HPLC applications. Information on column operation and maintenance is found in Chapter 5. A glossary of basic terminology is found in Chapter 2. Examples of HPLC applications are found in Chapters 6 and 7. HPLC of large biomolecules is discussed in Chapter 7.

3.2 GENERAL COLUMN DESCRIPTION AND CHARACTERISTICS

A typical HPLC column is a stainless steel tube filled with small-particle packing used for sample separation. HPLC columns can be categorized in a number of ways:

- By Column Hardware: Standard or cartridge, stainless steel, PEEK, titanium
- By Chromatographic Modes: Normal-phase (NPC), reversed-phase (RPC), ion-exchange (IEC), size-exclusion (SEC)

- By Dimensions: Prep, semi-prep, analytical, Fast LC, micro, nano
- By Support Types: Silica, polymer, zirconia, hybrid

The general characteristics of a typical HPLC analytical column in use today are:

- 50–250 mm long × 2.0–4.6 mm i.d. packed with a 3- or 5-μm, C18 silica-based bonded phase
- Flow rate range of 0.5–2 mL/min using a mobile phase of methanol or acetonitrile in water or aqueous buffer
- Sample loading of 1 ng to 1 mg
- Operating pressure ranges 500–3,000 psi
- Has 4,000–20,000 theoretical plates
- Costs $150–$450
- Typical column life time of 500–2,000 injections or 3–24 months

3.2.1 Column Hardware—Standard vs. Cartridge Format

Figure 3.1 shows a picture of HPLC columns of various sizes—from large preparative to smaller analytical columns. Figure 3.2a shows a schematic of a column inlet and end-fitting of a "standard" column. A typical analytical column is made from a stainless steel tube of ¼″ o.d., with end-fittings allowing connections to 1/16″ o.d. tubing (see Chapter 5 on column connections).

Figure 3.1. Picture of HPLC of various sizes (preparative, semi-preparative and analytical columns). Picture courtesy of Inertsil.com.

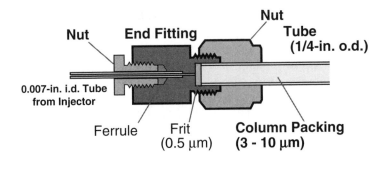

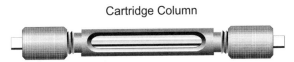

Cartridge Column

Figure 3.2. (a) Schematic of the inlet of a standard HPLC column showing details on the inlet 1/16″ o.d. tube with compression screw and ferrule, the end-fitting, and the inlet 0.5-μm frit. (b) Diagram of a PE-Brownlee cartridge column placed inside an aluminum cartridge holder with finger-tight end-fittings. This was one of the first cartridge column designs. Diagram courtesy of PerkinElmer, Inc.

The packing is held in place with a set of 0.5- or 2.0-μm porous end frits. For bio-separation or ion-chromatography, the stainless steel column hardware can be replaced by titanium or polyetheretherketone (PEEK) for more corrosion resistance. Micro LC columns (<0.5 mm i.d.) are packed in fused silica capillaries coated with polyamide for flexibility. Figure 3.2b shows a cartridge column design in which a column with two pressed-in frits is placed inside a cartridge holder. This design has a cost advantage since the actual column has no end-fittings, though a cartridge holder must be purchased for a first-time user. While most columns in the market are standard columns, some vendors offer columns in both standard and cartridge formats.

3.3 COLUMN TYPES

3.3.1 Types Based on Chromatographic Modes

HPLC columns can be categorized into four major modes (NPC, RPC, IEC, and SEC) in addition to other specialized modes (e.g., affinity, chiral, or specified applications).[1,3,9] Since reversed-phase chromatography (RPC) is used in 70–80% of all HPLC applications, RPC columns for small molecules are the focus of this discussion.

Table 3.1. HPLC Column Types Based on Inner Diameters (i.d.)

Column type*	Typical i.d. (mm)*	Typical flow rate (mL/min)	Void volume (mL)	Typical sample loading
Preparative	>25	>30	>50.0	>30 mg**
Semi-prep	10	5	7.5	5 mg**
Conventional	4.6	1	1.6	100 μg
Mini-bore or solvent saver	3.0	0.4	0.7	40 μg
Narrowbore	2.0	0.2	0.3	20 μg
Microbore	1.0	0.05	0.075	5 μg
Micro LC—Capillary	<0.5	10 μL/min	20 μL	1 μg
Nano LC—Capillary	<0.1	0.5 μL/min	1 μL	0.05 μg

*Designations of these column types are not universally accepted and may vary with manufacturers. Column i.d. is typical. Void volumes are based on lengths of 150 mm.
**Sample loading for preparative applications under overload conditions.

3.3.2 Types Based on Dimensions

Column dimensions—length (L) and column inner diameter (d_c or i.d.)—control column performance (N, speed, sensitivity, sample capacity) and its operating characteristics (flow rate, back pressure). Designations of various column types based on column inner diameters and their associated characteristics are shown in Table 3.1. Note that void volume, sample capacity, and operating flow rate are proportional to $(d_c)^2$, while detection limit, or sensitivity, is inversely proportional to $(d_c)^2$. Note also that prep columns (>10 mm i.d.), microbore (<1 mm i.d.), and micro columns (<0.5 mm i.d.) will require specialized HPLC instruments (see Chapter 4). There is a definitive trend toward the increased use of shorter and smaller inner diameter analytical columns due to their higher sensitivity performance and lower solvent usage.[9–11] This trend will be explored later.

3.3.3 Column Length (L)

Column length (L) determines column efficiency (N), analysis speed, and pressure drop (see Chapter 2):

Plate number (N) = L/H   (Since plate height H is a constant at a given flow rate, N is proportional to L.)

Figure 3.3 shows five chromatograms of a test mixture on columns of various lengths (L = 15, 30, 50, 75, and 150 mm), illustrating the effect of L on N, analysis time, and peak resolution. Note that longer columns have higher plate numbers and yield better resolution with longer analysis time. Note that

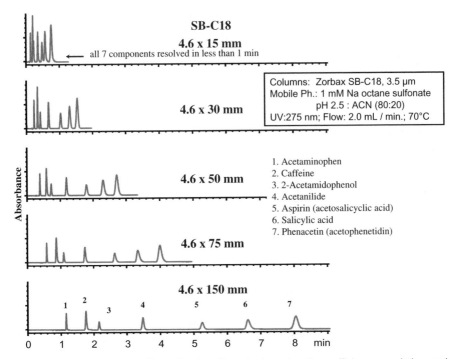

Figure 3.3. Chromatograms illustrating the effect of column length on efficiency, resolution, and sensitivity. Chromatograms courtesy of Agilent Technologies.

column pressure drop is also proportional to L. For simple sample mixtures, shorter columns might yield sufficient resolution with faster analysis times. This can represent a significant increase in productivity of 3–5 fold for many laboratories. Since plate numbers are additive, columns can be connected together to produce higher efficiency at the expense of longer analysis time and lower sensitivity due to more sample dilution. Typical column length ranges from 50 to 250 mm, though the use of shorter Fast LC columns (15–50 mm) is increasing.

3.4 COLUMN PACKING CHARACTERISTICS

The nature and characteristic of column packing containing the stationary phase is critical to the column performance and success of the intended applications.[1-3] Common types and characteristics of column packings in use today for small molecule separations are summarized below. Details are discussed further in this section.

- Support Type
  - Silica, polymer, hybrid

- Bonded Groups
  - C18, C8, C4, C3, amino, cyano, phenyl, polar-embedded (for RPC)
  - Diethylaminoethyl, sulfonate, quaternary ammonium, carboxylate (for IEC)
- Particle Size (d_p)
  - 2, 3, 5 µm (>10 µm for prep)
- Pore Size (d_{pore})
  - 60–200 Å
- Surface Area
  - 100–500 m²/g
- Ligand (Bonded-Phase) Density
  - 2–4 µmole/m²

3.4.1 Support Type

Silica (SiO_2) is the dominant support material, with excellent physical and chromatographic performance.[1,5] Columns packed with unbonded silica are rarely used for analytical purposes due to the strong adsorptive characteristics. Silanol groups (Si-OH) found on silica surfaces are typically bonded with monochlorosilanes to create a hydrophobic "liquid-like" stationary phase for reversed-phase applications.[1,12] Unreacted or residual silanols remaining after the bonding step are further reacted with a smaller silane (end-capped) to reduce the number of these adsorptive sites (Figure 3.4). One limitation of

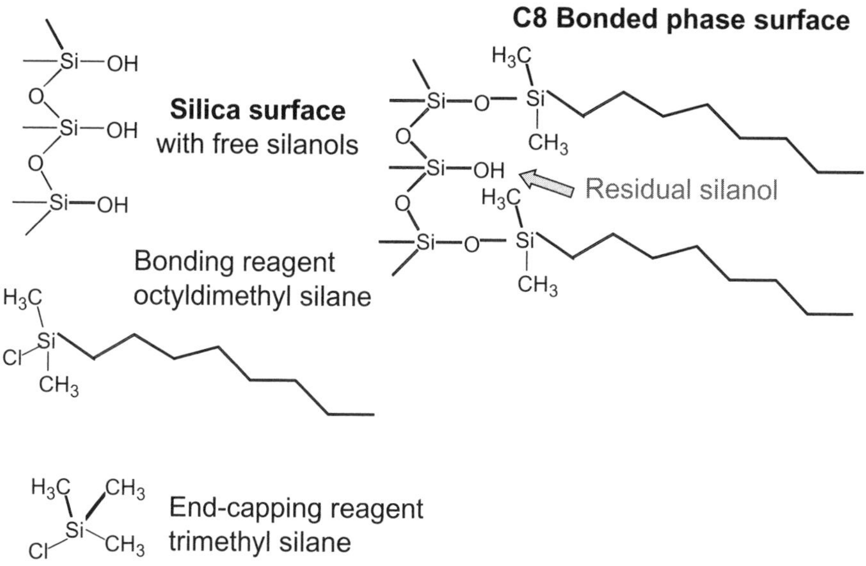

Figure 3.4. Structures of the typical bonding and end-capping reagents and diagram of the bonded silica with residual silanols.

silica-based bonded phases is the operating pH range of 2–8, which has been extended by recent innovations in bonding and support chemistries.

Polymer support materials such as cross-linked polystyrene-divinylbenzene, polyethers, and polymethacrylates have been used successfully for many years, mostly in bioseparations[1,3] and support for ion-exchange chromatography. Their strength and performance has improved in recent years, though they still lag behind silica in efficiency. Major advantages are wider pH range (1–14) and an absence of active silanols groups. Other materials used include alumina (mostly for NPC or preparative use) and zirconia (ZiO_2),[1,13] which can be coated or bonded to yield stable packing with unique selectivity and high-temperature stability.

3.4.2   Particle Size (d_p)

Particle size and size distribution define the quality of the support material and are the key determinants of efficiency and back-pressure of the column.[1–3] The effect of d_p on H is discussed in Chapter 2. For a well-packed column, H_{min} is approximated to 2–2.5 d_p. Also, since the van Deemter equation C term is proportional to d_p^2, columns packed with small particles have much less efficiency loss at high flow rates.[14] However, since column back-pressure is inversely proportional to d_p^2, columns packed with sub-3-μm particles can easily exceed the pressure limit of most HPLC instruments at 6,000 psi. Note that decreasing particle size while keep the L constant can increase column efficiency and peak resolution (Figure 3.5A) and also increase peak height and sensitivity (Figure 3.5B).

3.4.3   Surface Area and Pore Size (d_{pore})

Most chromatographic supports are porous to provide more surface area to maximize interaction of the solutes with the bonded stationary phase. Packings for small molecule separations have pore sizes ranging from 60 to 200 Å and surface areas of 100–500 m^2/g. In general, high surface area supports can yield bonded phase with higher density and thus higher retention (Figure 3.6). Note that small-pore packings are problematic for large biomolecules, which can become entangled or trapped in the pores leading to slower mass transfer and additional band broadening (see examples in Chapter 7 on column requirements for bio-separations).

3.4.4   Bonding Chemistries

Figure 3.4 shows the traditional bonding chemistry on a silica support by reacting the surface silanols with a bonding reagent such as dimethlyoctylchlorosilane to form a layer of hydrophobic stationary phase.[1–3,12] Since steric hindrance only allows ~50% bonding efficiency, the remaining residual silanols are further reacted or "end-capped" with a smaller silane such as

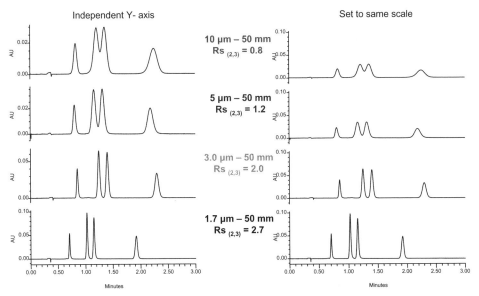

Figure 3.5. Chromatogram illustrating the effect of decreasing particle size (while keeping L = 50 mm) on efficiency, resolution, and sensitivity. Chromatograms courtesy of Waters Corporation.

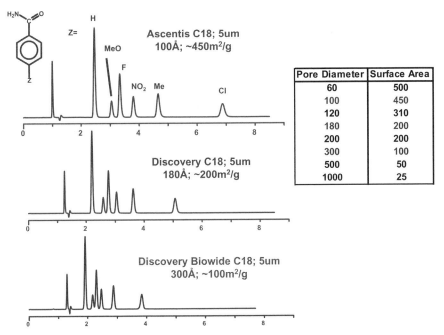

Pore Diameter	Surface Area
60	500
100	450
120	310
180	200
200	200
300	100
500	50
1000	25

Figure 3.6. Examples illustrating the increase of retention and resolution with increasing surface area of the silica support. Diagram courtesy of Supelco Inc.

Reversed-Phase (RPC)

C18	Octadecyl	$Si(CH_3)_2- (CH_2)_{17} - CH_3$
C8	Octyl	$Si(CH_3)_2- CH_2-CH_2- CH_2- CH_2- CH_2- CH_2- CH_2- CH_3$
CN	Cyano	$Si(CH_3)_2- CH_2- CH_2- CH_2- CN$
ϕ	Phenyl	$Si(CH_3)_2- C_6H_6$
Polar-embedded (Amide)		$Si(CH_3)_2- (CH_2)_3NHCO(CH_2)_{14}CH_3$

Normal-Phase (NPC)

Si	Silica	$Si - OH$
NH_2	Amino	$Si(CH_3)_2- CH_2- CH_2- CH_2- NH_2$

Ion-Exchange (IEC)

SP	Sulfo propyl	$Si(CH_3)_2 - CH_2- CH_2- CH_2- SO_3^-$
CM	Carboxymethyl	$Si(CH_3)_2-CO_2^-$
DEAE	Diethylaminoethyl	$Si(CH_3)_2 - CH_2-CH_2-CH_2-NH^+ (C_2H_5)_2$
SAX	Triethylaminopropyl	$Si(CH_3)_2-CH_2-CH_2-CH_2-N^+(C_2H_5)_3$

Figure 3.7. Structures of some common bonded phases.

trimethylchlorosilane. Some vendors might perform double or triple end-capping to reduce the residual silanol activity. Similarly, other bonding reagents can be used to yield a large variety of bonded phases for RPC, NPC, and IEC (Figure 3.7). Figure 3.8 shows the selectivity effect of five different bonded phases of the same silica support. New bonding chemistries to improve column performance are discussed in a later section.

3.4.5   Some General Guidelines for Bonded Phase Selection

Some general guidelines for selecting RP-bonded phases are summarized below.[6] Applications are described in Chapters 6 and 7.

- C18—Very hydrophobic, retentive, and stable phase. First choice for most separations.
- C8—Preferred for lower organic mobile phase applications for more polar solutes. Note that C8 has similar selectivity as C18 but is much less retentive.
- CN—Less retentive and different selectivity than C8. Note that many CN phases are much less stable than C18.
- Phenyl—For medium-polarity components. Unique selectivity for aromatics.

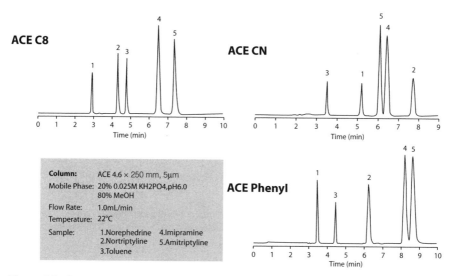

Figure 3.8. Comparative chromatograms illustrating the effect of selectivity of C8, cyano-, and phenyl-bonded phases. Diagram courtesy of MAC-MOD Analytical.

- C3 or C4—Less retentive than C8 or C18. Mostly used for protein separations on wide-pore supports.

3.5   MODERN HPLC COLUMN TRENDS

Understanding modern column trends is important since column technologies continue to evolve rapidly, resulting in new products with higher performance and more consistency.[9-11] Innovations have been targeting the traditional problem areas of silica-based RPC columns. They are:

- Batch-to-batch inconsistency
- Peak tailing for basic analytes
- Low column lifetime
- pH limitations (2–8)

Many of these problems have been resolved or minimized through developments in the last two decades. This section discusses key modern trends, including high-purity silica, hybrid particles, and novel bonding chemistries, and describes their impacts on column performance. The benefits of Fast LC and micro LC columns are also discussed with their salient applications.

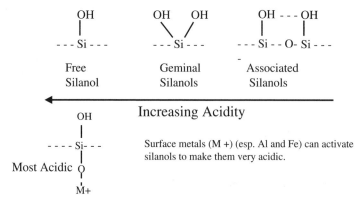

Figure 3.9. Schematic of various types of silanol and their relative acidities. Note that the silanol can be activated by an adjacent metal group. Adapted from and reprinted with permission from reference 2.

3.5.1 High-Purity Silica

One of the key advances in column technologies is the development of high-purity silica.[1,9] In recent years, it has become a *de facto* industry standard for almost all new column offerings. This development stems from the realization that batch-to-batch reproducibility and peak tailing of basic solutes are mostly caused by acidic residual silanols. Figure 3.9 shows different types of silanols and their relative acidity. The worst culprits turned out to be the very acidic silanols adjacent to and activated by metallic oxides. Many older silica materials have high metallic contents (e.g., Spherisorb) and are extremely acidic. They often require the use of amine additives in the mobile phase (e.g., tri-ethylamine) to prevent adsorptive interaction with basic analytes. The inherent variations of these active (acidic) silanols are also responsible for the lack of batch-to-batch consistency of these acidic silica materials.

The development of high-purity silica (99.995%) with <50 ppm of metals in the 1990s appeared to be one of the keys for dramatic improvements in batch-to-batch reproducibility and peak shapes of basic analytes. Other factors are improved bonding chemistries and quality control procedures. Figure 3.10 shows comparative chromatograms of two C18 columns and their differences in performance for basic analytes. Note that peak tailing of basic analytes is not caused by the number of residual silanols but rather the activity of these silanols. Figure 3.11 lists some popular RPC columns in use today, with most new products based on high-purity silica or hybrid particles.

3.5.2 Hybrid Particles

Another innovative approach to improve silica support is the development of hybrid particles.[3,16] In the silica synthesis of the first commercial hybrid

Waters Symmetry C18   *Hypersil BDS-C18*

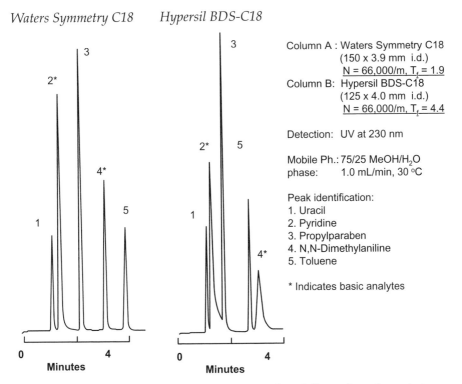

Column A : Waters Symmetry C18
            (150 x 3.9 mm i.d.)
            N = 66,000/m, T_f = 1.9
Column B: Hypersil BDS-C18
            (125 x 4.0 mm i.d.)
            N = 66,000/m, T_f = 4.4

Detection: UV at 230 nm

Mobile Ph.: 75/25 MeOH/H$_2$O
phase: 1.0 mL/min, 30 °C

Peak identification:
1. Uracil
2. Pyridine
3. Propylparaben
4. N,N-Dimethylaniline
5. Toluene

* Indicates basic analytes

Figure 3.10. Comparative chromatograms showing the effect of silica purity on the peak shapes of basic analytes. Note that the column on the right shows significant silanol activity causing considerable peak tailing due to its higher metallic contents in the base silica.

particles (Waters XTerra), tetraethoxysilane (the traditional monomer) is mixed with methyltriethoxysilane resulting in a hybrid silica support with 50% less surface silanols (Figure 3.12). Consequently, hybrid particles tend to have less residual silanol activity and less peak tailing for basic analytes. Another important advantage is the wider usable pH range of 2–12. Improved hybrid particles displaying even higher stability and efficiency performance have also been developed (Figure 3.11).

3.5.3 Novel Bonding Chemistries

The traditional bonding chemistry using a mono-functional silane achieves a maximum bonding efficiency of ~50% or a ligand density of ~4 μmole/m^2. These bonded phases suffer from several known disadvantages:

* At pH < 2, the Si-O bonds are subjected to acidic hydrolytic cleavage, causing the loss of the bonded phase.

- **Waters:** Symmetry, SunFire, *XTerra **, *ACQUITY**, *X-Bridge** Atlantis, NovaPak, μ-Bondapak, Spherisorb

- **Agilent:** Zorbax StableBond, Eclipse XDB, *Extend C18**, Bonus

- **Phenomenex:** *Luna**, Prodigy, *Synergi**, *Gemini**

- **Supelco:** Discovery, Ascentis, Supelcosil

- **Varian:** Inertsil, *Polaris**

- **Thermo:** HyPURITY, Hypersil, *Prism*, *Hypersil Gold **

- **MacMod:** ProntoSIL, ACE (Adv. Chrom. Tech.)

- **YMC:** YMCbasic, Pack Pro

- **Eka Chemicals:** Kromasil

- **GL Sciences:** Inertsil

- **Macherey Nagel:** Nucleosil

- **Merck KGaA:** Chromolith (Monolith)

- **Bischoff:** *ProntoSIL**

- **Grace:** Vydac, Platinum (Alltech)

- **Dionex:** Acclaim, Acclaim PA, *Acclaim PA2**

*Columns based on high-purity silica are <u>underlined</u>. Hybrid particles are **in bold**. Phases stable in high pH are italicized and marked with *.*

Figure 3.11. A partial list of popular silica-based columns including some recent new offerings. Columns based on high-purity silica are underlined. Columns packed with hybrid particles are in bold. Note that the ACQUITY (Waters) and Gemini (Phenomenex) are new second-generation hybrid columns.

- At pH > 8, the silica structure is prone to dissolution.
- Any unbonded acidic silanols might lead to peak tailing of basic analytes.

Several innovative bonding chemistries[1,9,17] have been developed to mitigate these problem areas:

- *Sterically Hindered Group*—Bulky di-isopropyl or di-isobutyl groups are incorporated in bonding reagents to protect the labile Si-O bonds against acid hydrolysis (see Figure 3.13).
- *Polyfunctional Silane Chemistry*—The use of di-functional or tri-functional silanes to create bonded groups with two or three attachment points leading to phases with higher stability in low or high pH, and lower bleed for LC/MS (see Figure 3.14a). Note that polyfunctional silane chemistry is more difficult to control.
- *Polar-Embedded Groups*—The incorporation of a polar group (e.g., carbamate, amide, urea, ether, sulfonamide) in the hydrophobic chain of the bonding reagent has led to a new class of popular phases with different

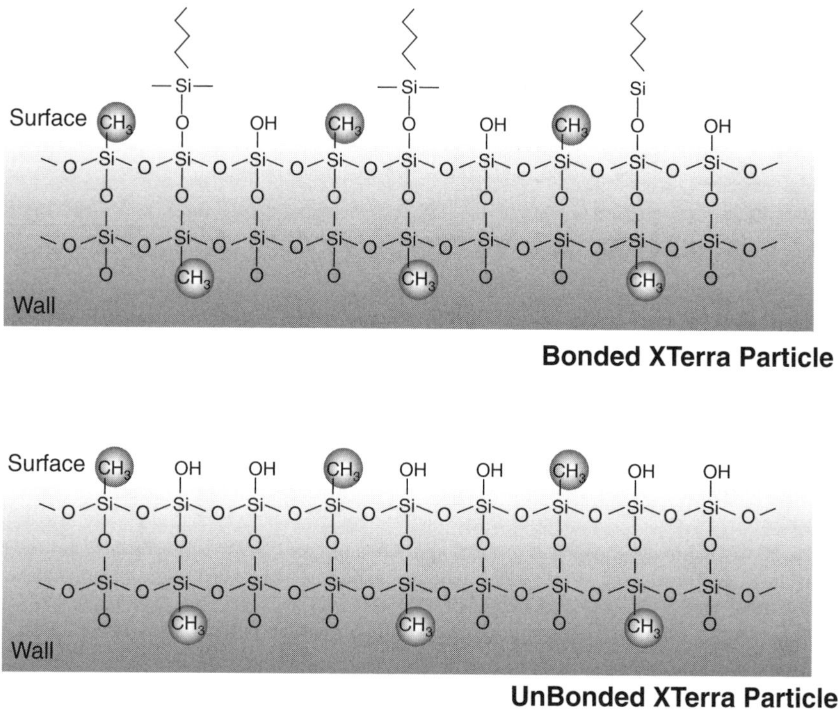

Bonded XTerra Particle

UnBonded XTerra Particle

Figure 3.12. Diagram illustrating the silica surface of unbonded and bonded Waters XTerra hybrid particles. Note that XTerra contains 50% less residual silanols by nature of its hybrid support chemistry. Diagram courtesy of Waters Corporation.

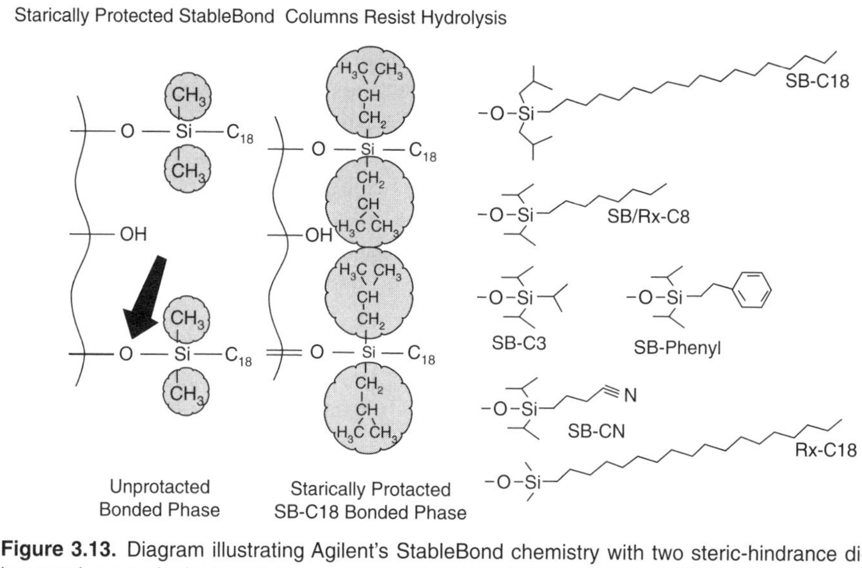

Starically Protected StableBond Columns Resist Hydrolysis

Unprotected Bonded Phase

Starically Protected SB-C18 Bonded Phase

Figure 3.13. Diagram illustrating Agilent's StableBond chemistry with two steric-hindrance di-isopropyl groups in the bonding reagent to protect the Si-O bond from acid hydrolytic cleavage. Diagram courtesy of Agilent Technologies.

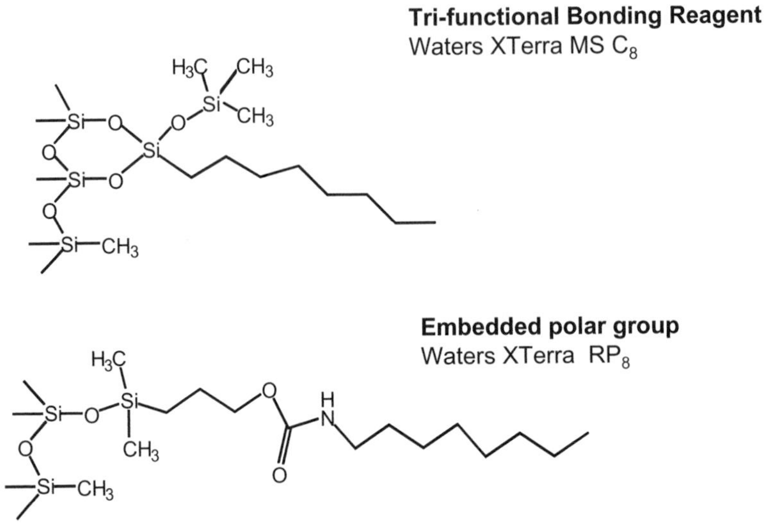

Tri-functional Bonding Reagent
Waters XTerra MS C$_8$

Embedded polar group
Waters XTerra RP$_8$

Figure 3.14. Diagrams showing the bonding chemistry used in Waters XTerra columns. (a) XTerra MS C8: The use of tri-functional bonding reagent with two or three attachment points leads to more acid resistance and less column bleed for LC/MS analysis. (b) XTerra RP8: The use of bonding reagent with an embedded-polar group (carbamate) leads to some shielding of the residual silanols as well as different selectivity due to additional hydrophilic interactions with the solutes. Diagram courtesy of Waters Corporation.

selectivity and better peak shape for basic analytes due to "shielding" of the silanols by the polar-embedded group (see Figures 3.14b and 3.15). These columns are generally less retentive than regular C18 or C8 counterparts for neutral and basic analytes. They are becoming quite popular for method development of difficult separations. More examples are shown in Chapter 8. Some examples of modern polar-embedded columns are Waters XTerra RP8 and RP18 (carbamate), Agilent Zorbax Bonus-RP (amide), Supelco Ascentis RP-Amide, Dionex Acclaim PA (sulfonamide) and PA2, Phenomenex Synergy Fusion-RP, ES Industries Chromegabond ODS-PI (urea), and Varian Polaris C18-ether. Figure 3.15 shows the column selectivity differences of several polar-embedded phases versus that of a conventional C18 phase.

Separation of very polar analytes is traditionally difficult in RPLC due to lack of retention of these compounds. Lowering the organic strength of the mobile phase to increase retention is effective only to about a few percentage points of organic modifier (e.g., 3–10% methanol). When the mobile phase is almost totally aqueous and the flow is stopped, sudden collapse of the hydrophobic bonded phase due to the "de-wetting" is often observed[1,2] (Figure 3.16). Polar-embedded phases are better suited for the separation of polar analytes since they are not prone to phase collapse due to repulsion of the polar-embedded

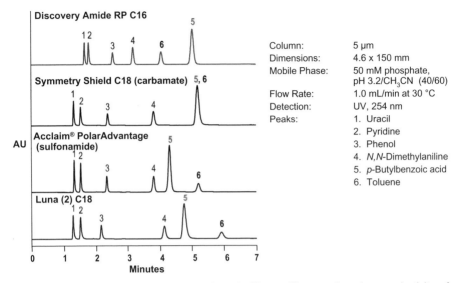

Column:	5 μm
Dimensions:	4.6 x 150 mm
Mobile Phase:	50 mM phosphate, pH 3.2/CH₃CN (40/60)
Flow Rate:	1.0 mL/min at 30 °C
Detection:	UV, 254 nm
Peaks:	1. Uracil
	2. Pyridine
	3. Phenol
	4. *N,N*-Dimethylaniline
	5. *p*-Butylbenzoic acid
	6. Toluene

Figure 3.15. Comparative diagram showing significant difference in column selectivity of various polar-embedded bonded phases versus that of Luna C18, a more conventional straight-chain C18 chemistry. Diagram courtesy of Dionex Corporation.

Column:	5 μm	Peaks:	1. Cytosine
Dimensions:	4.6 x 150 mm		2. Uracil
Mobile Phase:	2.5 mM MSA*		2. Thymine
Flow Rate:	1.0 mL/min at 30 °C	Protocol:	Each cycle consists of two steps:
Detection:	UV, 254 nm		1. Equilibrate column for 20 min before testing for 10 min
			2. Stop flow for 30 min before next cycle begins

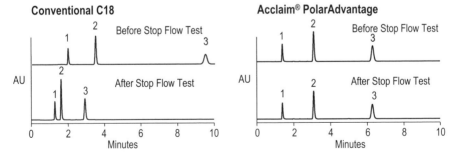

*Methanesulfonic acid

Figure 3.16. Chromatogram illustrating the phenomenon of phase collapse for convention C18 phases when used with highly aqueous mobile phases (chromatogram on the left-hand side). Polar-embedded phases are not prone to phase collapse and are better suited for the separation of very water-soluble analytes. Diagram courtesy of Dionex Corporation.

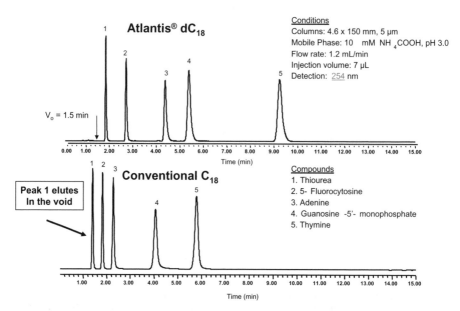

Figure 3.17. Comparative chromatograms of a low-coverage C18-bonded phase designed for the separation of polar water-soluble analytes versus that of a conventional high-coverage C18-bonded phase. Chromatograms courtesy of Waters Corporation.

group. Another approach to mitigate this phase collapse problem is to do a partial bonding (lower coverage) of a high-purity silica support to create a more hydrophilic bonded phase (Figure 3.17). This type of bonded phase is useful for separations of very water-soluble compounds and can be used with a 100% aqueous mobile phase. Examples are Alltech Platinum EPS and Waters Atlantis dC18.

3.5.4 Fast LC

The use of short columns (15–50 mm) packed with small particles (i.e., <3 μm) has several significant benefits in the analysis of simple mixtures and in high-throughput screening,[18–20] such as:

- Fast analysis—isocratic (1–3 min) or gradient (2–10 min)
- Rapid method development and validation
- Lower solvent consumption—50–80% less than conventional columns
- Increase mass sensitivity—2–5 fold

Today, Fast LC is the easiest way to increase HPLC laboratory productivity by 3–5 fold without sacrificing resolution, sensitivity, or assay reliability.[19] Figures 3.18–3.20 demonstrate the versatility and dramatic speed achievable with Fast LC. Chapter 4 discusses the effect of extracolumn band-broadening

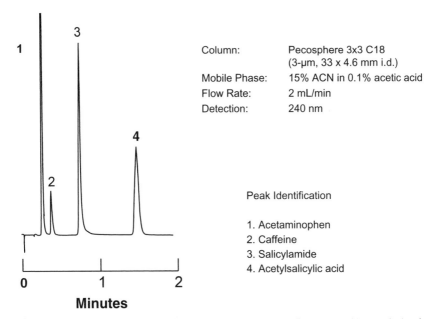

Column:	Pecosphere 3x3 C18 (3-µm, 33 x 4.6 mm i.d.)
Mobile Phase:	15% ACN in 0.1% acetic acid
Flow Rate:	2 mL/min
Detection:	240 nm

Peak Identification

1. Acetaminophen
2. Caffeine
3. Salicylamide
4. Acetylsalicylic acid

Figure 3.18. Fast LC separation of an analgesic tablet extract. Reprinted with permission from reference 19.

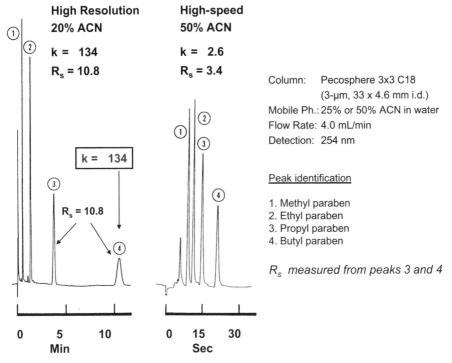

High Resolution
20% ACN

k = 134
R$_s$ = 10.8

High-speed
50% ACN

k = 2.6
R$_s$ = 3.4

Column:	Pecosphere 3x3 C18 (3-µm, 33 x 4.6 mm i.d.)
Mobile Ph.:	25% or 50% ACN in water
Flow Rate:	4.0 mL/min
Detection:	254 nm

Peak identification

1. Methyl paraben
2. Ethyl paraben
3. Propyl paraben
4. Butyl paraben

R$_s$ measured from peaks 3 and 4

Figure 3.19. Diagram illustrating the versatility of Fast LC in high-speed and high-resolution separation of four paraben antimicrobials. Diagram reprinted with permission from reference 19.

Column: Zorbax StableBond C18
                 (1.8-µm, 50x 4.6 mm i.d.)
Mobile Phase: A: water + 0.1%HCOOH
                 B: ACN + 0.1%HCOOH
Gradient: 50-100% B in 0.65 min @ 32 °C
Detection: 245 nm with a 500-nL PDA flow cell
Sample: Test mix of 9 alkylphenones

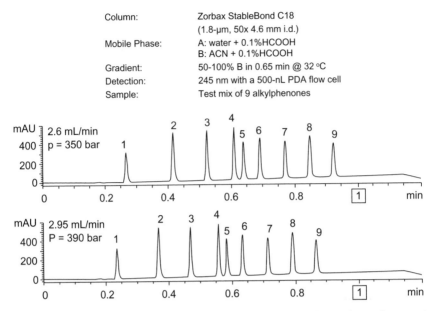

Figure 3.20. Sub-minute high-throughput screening application using a short column packed with 1.8-µm particles at high linear flow velocities and short gradient time. Diagram courtesy of Agilent Technologies.

and instrumental requirements. Figure 3.21 shows van Deemter curves, which illustrate how the sub-2-µm particles can further improve efficiencies particularly at higher flow rate due to their smaller C terms. These columns have much higher back-pressure (ΔP is proportional to $1/d_p^2$) and are more susceptible to extracolumn effects. Nevertheless, the recent availability of low-dispersion ultra-high-pressure LC instruments coupled with columns packed with sub-2-µm particles might set a new performance benchmark in HPLC technologies.[21,22]

3.5.5 Micro LC

The practice of micro LC using columns <1-mm i.d. such as microbore (1 mm), micro LC (<0.5-mm i.d.), and nano LC (<0.1-mm i.d.) has these distinct benefits[23]:

- Increased mass sensitivity due to lower sample dilution (Figure 3.22).
  - Useful for samples of limited availability (proteins/peptides, pharmacokinetic samples)
- Savings in solvent consumption and associated disposal costs.
  - Typical flow rates are 50 nL/min to 50 µL/min
- Better compatibility with LC/MS (e.g., electrospray interface).

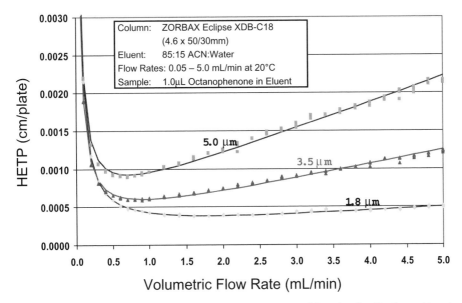

Figure 3.21. Van Deemter curves of 5-, 3.5-, and 1.8-µm particles showing the lower H and extreme flatness due to the smaller C term of the smaller particles. Diagram courtesy of Dr. Bill Barbers of Agilent Technologies.

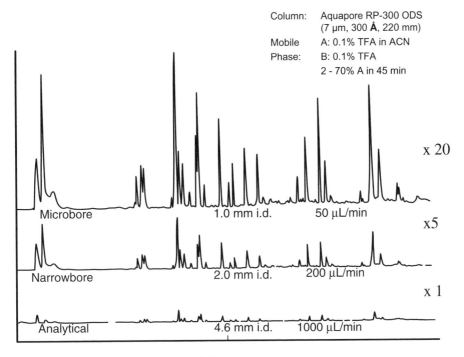

Figure 3.22. Comparative RPLC chromatograms of tryptic digest illustrating the sensitivity enhancement of columns with smaller inner diameters. Diagrams courtesy of PerkinElmer, Inc.

Current micro LC applications are primarily in micropurifications of proteins/peptides and in proteomics research. Chapter 4 discusses the stringent instrumentation requirements for micro and nano LC, which must have instruments with very low dispersion and dwell volumes.

3.5.6 Monoliths

Monolithic technology first originated as an alternate technique to fabricate capillary columns. It is a radical departure from packed-column technology and uses *in situ* polymerization to form a continuous bed of porous silica[24] inside the fused silica capillaries. Because end frits are problematic in packed capillaries, this new approach eliminates this problem since no end frits are required for monoliths. For years, capillary monoliths remained a scientific tool for academic research.

Surprisingly, the first commercial monolith, Chromolith, introduced by Merck EKaA, is an analytical sized column, available as 50- and 100-mm × 4.6-mm i.d.[25] The major advantage of Chromolith is the lower pressure drop, which is claimed by the manufacturer to have the performance of 3-μm particles and pressure drop of 5-μm particles (Figure 3.23). Attempts to manufacture smaller-diameter columns have met with considerable difficulties, particularly with the "cladding" or the external bonding of these preformed

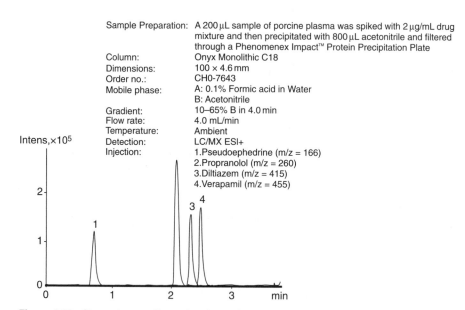

Figure 3.23. Chromatogram illustrating the performance of a monolith column in a bioanalytical application. Diagram courtesy of Phenomenex, Inc.

silica rods with PEEK, however, a 3-mm i.d. Chromolith is now available. Currently, the pressure limit of these commercial monoliths is ~3,000 psi. It remains to be seen in the next decade whether monolith technology can be competitive with advances in packed-column technologies.

3.6 GUARD COLUMNS

A guard column is a small column placed before the analytical column to protect it from particles or contaminants from the samples.[26] Ideally, guard columns are packed with the same materials as analytical columns and should not cause a significant increase in pressure or performance degradation (Figure 3.24). Guard columns are commonly used in applications involving "dirty samples"—environmental or bioanalytical samples where more extensive sample clean-up might not be feasible. Some users might prefer the use of an in-line filter, though it does not have the sample capacity as a well-designed guard column. Another type of guard column, termed a "scavenger column," is placed between the pump and the injector.[26] Its purpose is to protect the analytical column from mobile phase contaminants. Its use can be desirable for small particle columns, which are easily plugged, and in ion-pair chromatography to prefilter the mobile phase. Note that a scavenger column will increase the system dwell volume.

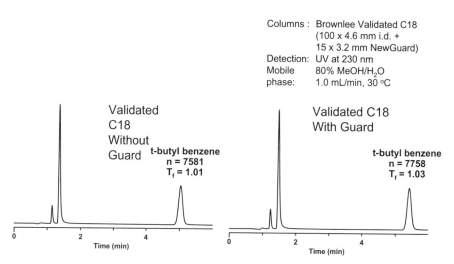

Columns : Brownlee Validated C18
(100 x 4.6 mm i.d. +
15 x 3.2 mm NewGuard)
Detection: UV at 230 nm
Mobile 80% MeOH/H$_2$O
phase: 1.0 mL/min, 30 °C

Validated C18 Without Guard
t-butyl benzene
n = 7581
T$_f$ = 1.01

Validated C18 With Guard
t-butyl benzene
n = 7758
T$_f$ = 1.03

Figure 3.24. Comparative chromatogram showing the performance with and without a guard column. Diagram courtesy of PerkinElmer, Inc.

3.7 SPECIALTY COLUMNS

3.7.1 Bioseparation Columns

The HPLC of large biomolecules such as proteins and DNA often requires specialized columns packed with wide-pore polymer or silica-based bonded phase with extra-low silanol activity.[12,15] Alternate approaches are pellicular materials or very small nonporous particles. Some of these columns are packed in PEEK or titanium hardware to allow the use of high-salt mobile phase and to prevent possible protein denaturing by metallic leachates. Further details on bio-separations and application examples are discussed in Chapter 7.

3.7.2 Chiral Columns

Chiral columns are packed with stereo-specific sorbents for the separation of stereoisomers in the sample.[27] Many chiral columns operate in hydrophilic interaction mode (e.g., Pirkle-type) whereas others are used in reversed-phase mode (proteins, macrocyclic antibiotics, polysaccharides, etc.). The use of these columns is critical in the development of chiral drugs. Most chiral columns are quite expensive and many older chiral columns have low efficiencies and limited lifetimes. Examples of chiral separations are shown in Chapter 6. Alternately, convention columns can also be used for chiral separations using a chiral selector in the mobile phase.[27]

3.7.3 Application-Specific Columns

While most columns are general-purpose, a number of columns are marketed for specific applications. Examples are columns for environmental analysis (carbamates, polynuclear aromatic hydrocarbons) or food testing (amino acids, organic acids, sugars). These columns are often shipped with chromatograms demonstrating the performance of the specific application. More examples of specific applications and GPC columns for polymer characterization are described in Chapter 7.

3.8 COLUMN SELECTION GUIDES

Column selection during method development often reflects personal preferences or prior experience.[1,4,6] Nevertheless, some general guidelines can be suggested based on consensus of experienced chromatographers. Note that these recommendations focus on RPLC.

- Select columns packed with 3- or 5-μm high-purity silica-bonded phases from a reputable manufacturer.
  - 50–100 mm × 4.6 mm for simple samples (e.g., assays and dissolution)
  - 50–150 mm × 3.0–4.6 mm for purity testing or more complex samples

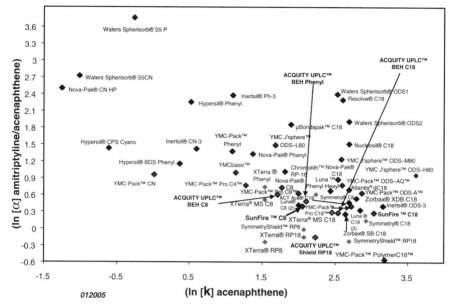

Figure 3.25. RPLC column selectivity chart. X-axis is an indication of the hybrophobicity and Y-axis is an indication of the silanophilic activity of the bonded phase. This comparative chart is useful for selecting equivalent or dissimilar columns from various vendors. Diagram courtesy of Waters Corporation.

- 20–150 mm × 2.0 mm columns for LC/MS
- 10–50 mm × 2.0 mm columns for high-throughput screening (HTS)
- Explore selectivity differences between C18, C8, polar-embedded, cyano- or phenyl-bonded phases.
  - Consult column selectivity chart to select similar (equivalent) or dissimilar (with different selectivity) columns based on its hydrophobicity or silanol activity (Figure 3.25).
- For low-pH applications (pH < 2.0), select columns resistant to acid hydrolysis of the bonded groups.
- For high-pH applications (pH > 8), select columns stable at high pH (Figure 3.11).

One important piece of advice for developing new methods or revalidating older methods is the selection of columns packed with high-purity silica support (see Figure 3.11) and novel bonding chemistries with wider pH ranges. These modern columns have improved reproducibility, stability, and versatility to ensure better assay consistency. Developing new methods using older columns just because they are at hand is not a good economical decision in the long run.

Table 3.2 shows a compiled listing of some common HPLC sorbents and their characteristics (surface area and pore size). Note that the retentive

Table 3.2. Listing of Common HPLC Sorbents and Attributes

Manufacturer	Sorbent	Surface Area (m^2/g)	Pore Size A	Notes
Agilent	Zorbax	300	70	
Agilent	Zorbax 300SB	45	300	
Agilent	Zorbax-Rx	180	80	
Bischoff	ProntoSIL	100	300	
Eka Chemicals	Kromasil	340	100	
GL Sciences	Inertsil ODS2	320	150	
GL Sciences	Inertsil ODS3	450	100	
Grace	Platinum (Alltech)	200	100	
Grace	Prevail (Alltech)	300	110	
Grace	Vydac 201-TP	90	300	
Hichrom	ACE	300	100	
Hamilton	PRP-1	415	100	Polymer
Macherey Nagel	Nucleosil 100	350	100	
Macherey Nagel	Nucleosil 120	200	120	
Merck KGaA	Chromolith	300	130	Monolith
Merck KGaA	LiChrospher	350	100	
PerkinElmer	Pecosphere	170	80	
Phenomenex	Gemini	375	110	Hybrid
Phenomenex	Luna	400	100	
Phenomenex	Prodigy	310	150	
Phenomenex	Synergi	475	80	
Polymer Labs	PLRP-S 100	510	100	Polymer
Restek	Allure	500	60	
Restek	Ultra	300	100	
Supelco	Ascentis	450	100	
Supelco	Discovery	200	180	
Supelco	Supelcosil	170	120	
Thermo	Betasil	310	100	
Thermo	BetaBasic	200	150	
Thermo	Hypersil	170	120	
Thermo	Hypersil Gold	220	175	
Thermo	HyPURITY	200	180	
Varian	Polaris	180	180	
Waters	ACQUITY (X-Bridge)	185	135	Hybrid
Waters	Atlantis	340	90	
Waters	μ-Bondapak	330	125	
Waters	NovaPak	120	60	
Waters	Spherisorb	220	80	
Waters	SunFire	340	100	
Waters	Symmetry	335	100	
Waters	Symmetry 300	110	300	
Waters	XTerra	175	125	Hybrid
YMC	J'sphere	510	80	

Column Match: Select a Column

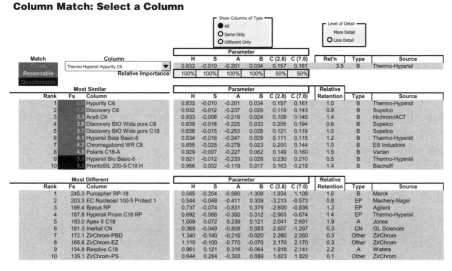

Figure 3.26. Results from Column Match, a database program allowing the user to select a similar or dissimilar column to that of the specified column in the existing method.

characteristics of the sorbent are dependent mostly on its surface area, ligand density and bonding type. More detailed specifications are available directly from the manufacturers.

Figure 3.25 shows a column selectivity chart of many popular reversed-phase bonded phases based on studies by Neue et al. The hybrophobicity of the sorbent is measured by the log k of a nonpolar solute (acenaphthene) and plotted in the X-axis. The silanophilic activity as indicated by the log α of amitriptyline versus aceanaphthene, which is plotted in the Y-axis. This comparative chart is useful for selecting equivalent or dissimilar bonded phases or columns from various vendors. The interested reader is referred to other extensive research work on the chromatographic classification and comparison of commercially available reversed-phase columns.[28,29] In an extensive study of more than 200 columns, Snyder, Dolan, and Carr[29] used five parameters (hydrophobicity (H), steric resistance (S*), hydrogen bonding acidity and basicity (A, B), and cation exchange (C)) to categorize column selectivity. A database program (Column Match) is commercially available from Rheodyne that allows users to select either similar or different columns to a column specified in the current method.

3.9 SUMMARY

This chapter provides an overview of essential column concepts, including a general description, types, packing, modern trends, and selection guide. While

the most popular column in use today is a 100–250 mm long, 4.6-mm i.d. column packed with 5-μm C8 or C18 silica-based bonded phase, there is a strong trend toward shorter or smaller inner diameter columns packed with 3-μm or <2-μm particles with novel bonding chemistries. The dominance of silica as the primary support material appears unchallenged and significant performance enhancements have been made possible with high-purity silica, hybrid particles with wider pH applications, and novel bonding technologies. The reader is urged to take advantage of these significant advances by developing more consistent and higher-performance HPLC methods using these improved modern columns.

3.10 REFERENCES

1. U.D. Neue, *HPLC Columns: Theory, Technology, and Practice*, Wiley-VCH, New York, 1997.
2. L.R. Snyder, J.J. Kirkland, and J.L. Glajch, *Practical HPLC Method Development*, 2nd Edition, Wiley-Interscience, New York, 1997.
3. K. Unger, *Packings and Stationary Phases in Chromatographic Techniques (Chromatographic Science)*, Marcel Dekker, New York, 1990.
4. S. Ahuja and M.W. Dong (eds.), *Handbook of Pharmaceutical Analysis by HPLC*, Elsevier, Amsterdam, 2005.
5. K. Unger, Porous Silica, Journal of Chromatography Library, Volume 16, Elsevier Scientific Publishing Co., New York, 1979.
6. L.R. Snyder and J.J. Kirkland, *Introduction to Modern Liquid Chromatography*, *2nd Edition*, John Wiley & Sons, New York, 1979.
7. V.R. Meyer, *Practical HPLC*, 4th edition, John Wiley, New York, 2004.
8. W.R. LaCourse, *Anal. Chem.*, **72**, 2813 (2000); **74**, 37R (2002).
9. R.E. Majors, *LC.GC. Eur.*, **16(6a)**, 8 (2003).
10. R.E. Majors, *LC.GC.*, **22(9)**, 870 (2004).
11. R.E. Majors, *LC.GC.*, **22(3)**, 230 (2004).
12. U. Neue, in R.A. Meyer, ed., Encyclopedia of Analytical Chemistry, John Wiley & Sons, Chichester, 2000, pp. 11450–11472.
13. C. Dunlap, C.V. McNeff, D. Stoll, and P.W. Carr, *Anal. Chem.*, **73(21)**, 591A (2001).
14. J.J. Van Deemter, F.J. Zuiderweg, and A. Klinkenberg, *Chem. Eng. Sci.*, **5**, 271 (1956).
15. R.L. Cunico, K.M. Gooding, and T. Wehr, *Basic HPLC and CE of Biomolecules*, Bay Bioanalytical Laboratory, Richmond, CA, 1998.
16. K.D. Wyndham, et al., *Anal. Chem.*, **75(24)**, 6781 (2003).
17. J.J. Kirkland, J.B. Adams Jr., M.A. van Straten, and H.A. Claessens, *Anal. Chem.*, **70**, 4344 (1998).
18. M.W. Dong, *Today's Chemist at Work*, **9(2)**, 46 (2000).
19. M.W. Dong and J.R. Gant, *LC.GC.*, **2(4)**, 294 (1984).
20. U.D. Neue and J.R. Mazzeo, *J. Sep. Sci.*, **24**, 921 (2001).
21. L. Tolley, J.W. Jorgenson, and M.A. Moseley, *Anal. Chem.*, **73(13)**, 2985 (2001).

22. D. Jerkovich, J.S. Mellors, and J.W. Jorgenson, *LC.GC.*, **21(7)**, 600 (2003).

23. D. Ishii, *Introduction to Microscale High-Performance Liquid Chromatography*, VCH Publishers Inc., New York, 1988.

24. N. Tanaka, H. Kobayashi, K. Nakanishi, H. Minakuchi, and N. Ishizuka, *Anal. Chem.*, **73**, 420A (2001).

25. K. Cabrera, D. Lubda, H.M. Eggenweiler, H. Minakuchi, and K.J. Nakanishi, *High Resol. Chromatogr.*, **23**, 93 (2000).

26. M.W. Dong, J.R. Gant, and P.A. Perrone, *LC.GC.*, **3**, 786 (1985).

27. H.Y. Aboul-Enein and I. Ali, *Chiral Separations by Liquid Chromatography: Theory and Applications* (Chromatographic Science, Volume 90), Marcel Dekker, New York, 2003.

28. M.R. Euerby and P. Petersson, *J. Chromatogr.*, **994**, 13 (2003).

29. L.R. Snyder, J.W. Dolan, and P.W. Carr, *J. Chromatogr.*, **1060**, 77 (2004).

3.11 INTERNET RESOURCES

http://www.chem.agilent.com
http://www.alltechweb.com
http://www.dionex.com
http://www.esind.com
http://www.gls.co.jp
http://www.hamiltoncompany.com
https://www.macherey-nagel.com
http://www.mac-mod.com/
http://www.merck.de
http://www.microlc.com
http://www.microsolvtech.com/
http://www.phenomenex.com
http://instruments.perkinelmer.com
http://www.polymerlabs.com
http://www.restekcorp.com/h
http://www.sigmaaldrich.com/Brands/Supelco_Home.html
http://www.thermo.com
http://www.tosoh.com/
http://www.varianinc.com
http://www.vydac.com/
http://www.waters.com
http://www.zirchrom.com/

4

HPLC INSTRUMENTATION AND TRENDS

Modern HPLC for Practicing Scientists, by Michael W. Dong
Copyright © 2006 John Wiley & Sons, Inc.

4.1 INTRODUCTION

4.1.1 Scope

High-performance liquid chromatography (HPLC) is a versatile analytical technique using sophisticated equipment refined over several decades. An in-depth understanding of the working principles and trends is useful for more effective application of the technique. This chapter provides the reader with a concise overview of HPLC instrumentation, operating principles, recent advances, and modern trends. The focus is on the analytical scale HPLC systems and modules (pump, injector, and detectors). System dwell volume and instrumental bandwidth are discussed, with their impact on small-diameter column applications. Specialized HPLC systems are described, as are their particular attributes. Guidelines for selecting HPLC equipment and data systems are reviewed. Readers are referred to books,[1-11] review articles,[12,13] training software,[14] and manufacturer literature and Web sites for additional details.

4.1.2   HPLC Systems and Modules

Today's liquid chromatographs have excellent performance and reliability. A typical HPLC system consists of a pump, an injector, a column, a detector, and a data-handling device (see schematic diagram in Figure 4.1a). Figure 4.1b is a pictorial representation of an earlier HPLC system comprising of an isocratic pump, a manual injector, a UV detector, and a PC-based data system. These simpler systems were prevalent in the 1980s. Modern HPLC systems are likely to resemble those in Figure 4.3, consisting of a multisolvent pump, an autosampler, an online degasser, a column oven, a UV/visible or a photo diode array detector, and a data-handling workstation, which also controls the entire HPLC system.

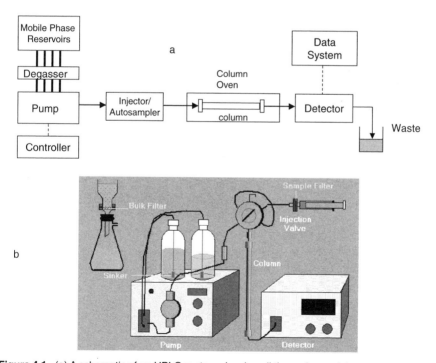

Figure 4.1. (a) A schematic of an HPLC system showing all the major modules or components. (b) A graphic representation of an earlier isocratic HPLC system. Reprinted with permission from Academy Savant.

a. Agilent Series 1100

c. Waters Alliance

b. JASCO L-P Mixing System

d. Shimadzu 2010

 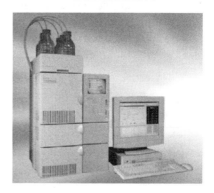

Figure 4.2. Examples of modular and integrated HPLC systems. Modular systems: (a) Agilent 1100 Series, (b) Jasco low-pressure mixing system. Integrated systems: (c) Waters Alliance, (d) Shimadzu 2010.

HPLC systems can be either modular or integrated. Figures 4.2a and 4.2b show modular systems consisting of separate but stackable modules. A modular system is usually more serviceable because each separate module can be exchanged if it malfunctions. Figures 4.3c and 4.3d show integrated systems where modules are built inside single housings, which tend to have cleaner looks. These built-in modules are controlled by a single control board inside the housing and cannot function as separate units outside the system. Integrated systems supposedly have better module integration (how the modules work together to yield better overall performance). In practice, both types are popular, with little noticeable performance differences to the user.

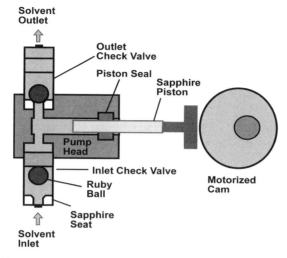

Figure 4.3. A schematic of a reciprocating single-piston pump.

4.2 HPLC SOLVENT DELIVERY SYSTEMS

A modern HPLC solvent delivery system consists of one or more pumps, solvent reservoirs, and a degassing system. Typical requirements of an analytical HPLC pumps are:

- Provide precise and pulse-free delivery of solvents at typical flow rates range of 0.1–10 mL/min and pressure up to 6,000 psi (42 MPa)
- Compatible with common organic solvents, buffers, and salts
- Reliable operation with long pump seal life
- Accurately blend solvents and generate gradient profiles (multisolvent pumps only)
- Easy to maintain and service

Figure 4.3 shows the schematic of a reciprocating pump mechanism used in most HPLC pumps. Here, a motorized cam drives a piston to deliver solvent through a set of check valves. A microprocessor coordinates the piston speed with other components. Since only the inward piston stroke delivers the liquid, a pulse dampener is used to reduce flow fluctuations. All components in the fluidic path are made from inert materials (e.g., stainless steel pump heads, ruby balls and sapphire seats in check valves, sapphire pistons, and fluorocarbon pump seals).

This simple reciprocating pump design has undergone numerous innovations, as summarized in Table 4.1. Note that gradient pump performance at lower flow rates (<200 μL/min) is particularly difficult to achieve due to a stan-

Table 4.1. Innovations for Enhancing Pump Performance

Pump performance	Innovations
Piston seal life	Spring-loaded seal, self-aligning piston, piston seal wash
Blending accuracy	Dual piston in-series design, high-speed proportioning valve, auto compensation for compressibility, and ΔV_{mixing}
Reduced pulsation	Rapid-refilling pump mechanism, pulse dampener
Better low-flow performance	Micro-pistons, variable stroke length, active check valve, syringe pump
Biocompatibility	Titanium or PEEK-based system (to lessen corrosion by salt or the leaching of metal ions from the system)
Multi-solvent pump	Low-pressure mixing system, online degasser
Reduced dwell volume	High-pressure mixing system, low-volume mixer

dard piston size of 100 μL. Innovations such as variable pump stroke mechanisms and micro pistons have extended acceptable low-flow performance down to 50–100 μL/min. For micro or nano LC, either flow-splitting or specialized pumping systems are required.

4.2.1 High-Pressure and Low-Pressure Mixing Designs in Multisolvent Pumps

Multisolvent pumping systems are classified by how solvent blending is achieved, as shown in Figure 4a and 4b. In low-pressure mixing designs typically found in quaternary pumps, a single pump draws mobile phases from a four-port proportioning valve (Figure 4.4b). The pump microprocessor controls the solvent composition of each intake piston stroke by the timed opening of each port. Note that solvent blending occurs inside the pump at low pressure. Effective degassing is mandatory in this design to prevent outgassing during blending. In a low-pressure mixing design, the pump head, pulse dampener, drain valve, filter, and pressure transducer all contribute to the internal liquid hold-up volumes or dwell volumes. However, a major advantage of low-pressure mixing pumps is the simplicity of using a single pump for multisolvent applications.

In high-pressure mixing systems, two or more separate pumps are used to mix solvents at high pressures (Figure 4.4a). A separate controller is often needed to change the flow rates of each pump to generate different blends or gradient profiles. An external mixer is often required. The disadvantage of high-pressure mixing systems is the cost and maintenance requirements of two or more pumps in a multisolvent system. The big advantage is the inherently lower dwell volume for applications using small-diameter columns. In modern binary high-pressure mixing systems, each pump is connected to two solvent

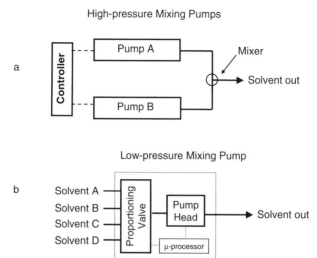

Figure 4.4. Schematic diagrams of a low-pressure mixing system (a) using a single pump and a four-port proportioning valve, and a high-pressure mixing system (b) using two separate pumps and a controller to blend solvents under high pressure.

reservoirs through an automated valve to increase solvent selection to four solvents.

4.2.2 System Dwell Volume

Also known as gradient delay volume, system dwell volume is the liquid hold-up volume of the HPLC system from the point of solvent mixing to the head of the column. This includes the additive volumes of the injector, the sample loop, all fluidic connection tubing, and any internal pump volumes of a low-pressure mixing system. The typical dwell volume of a modern HPLC is 0.5–2 mL, but can be as high as 5–7 mL in older systems.

High-pressure mixing systems have lower dwell volumes because the point of mixing is external to the pumps. Note that dwell volume is inconsequential in isocratic analysis but becomes important in gradient analyses because it adds gradient delay time to the analysis. It becomes critical for gradient applications at low flow rates. For instance, a dwell volume of 1 mL represents a 5-min gradient delay at 200 μL/min, used typically in LC/MS, or a 20-min delay at 50 μL/min for microbore (1-mm i.d.) columns. For these reasons, there is a revival of high-pressure mixing systems for high-throughput screening (HTS), LC/MS, and micro LC systems.

The system dwell volume can be measured using a UV detector as follows:

- Disconnect the column and replace it with a zero-dead-volume union.
- Place 0.5% acetone in water in reservoir B and water in reservoir A.

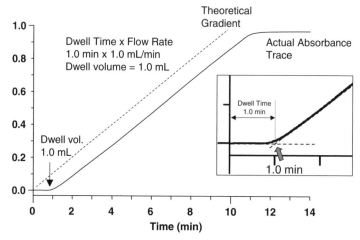

Figure 4.5. The gradient absorbance trace used to measure the system dwell volume of a Waters Alliance system. Inset shows the intersection point marking the gradient onset.

- Set a linear gradient program from 0% to 100% B in 10 min at a flow rate of 1.0 mL/min and a detection wavelength of 254 nm.
- Start the solvent gradient and the data system to record the detector signal.
- Measure the intersection point of the extrapolated absorbance trace with the baseline to record the dwell time, as shown in Figure 4.5. Multiply dwell time by flow rate to obtain the dwell volume.

4.2.3 Trends

Modern pumps have more features and better reliability and performance than earlier models because of better designs in seals, pistons, and check valves as well as innovations such as dual-piston in-series and piston seal wash.[2] Performance at low rates can be improved by variable stroke mechanism, micro pistons, or active check valves. The fluidic components in more inert pumps for bio-purification or ion-chromatography are often constructed from titanium or polyetheretherketone (PEEK). Low-pressure mixing quaternary pumps have become standard equipment in research laboratories whereas high-pressure mixing pumps are popular for LC/MS, HTS, and micro LC applications.

4.3 INJECTORS AND AUTOSAMPLERS

An HPLC injector is used to introduce the sample to the column under high pressure. A common injector is the Rheodyne model 7125 or 7725 injector,

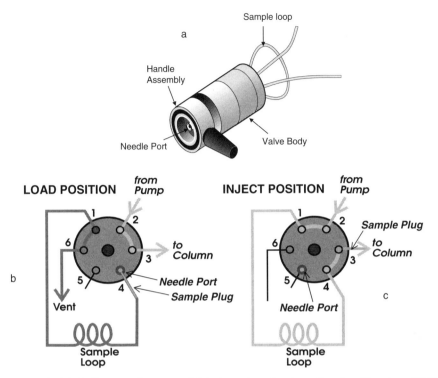

Figure 4.6. (a) A diagram of a manual injector valve. A schematic of a Rheodyne 7125 injector valve during the LOAD (b) and INJECT (c) cycle during a partial-loop injection operation.

which consists of a six-port valve with a rotor, a sample loop, and a needle port (Figure 4.6a). For manual injection, a syringe with a 22-gauge blunt-tip needle is used to introduce a precise sample aliquot into the sample loop at the LOAD position (Figure 4.6b). The sample is delivered into the column by switching the valve to the INJECT position (Figure 4.6c). For quantitative analysis, the sample aliquot should not exceed half of the sample loop volume.[2] Note that the sample is back-flushed into the column to minimize band-broadening in the sample loop.

4.3.1 Operating Principles of Autosamplers

An autosampler allows the automatic sample injection from sample vials or 96-well microplates (Figure 4.7a). An autosampler reduces labor cost and increases productivity and testing precision.

Figure 4.7b shows a schematic diagram of a typical "XYZ" autosampler consisting of a motorized injection valve and a moving sampling needle. The sampling needle assembly is mounted in a platform allowing movement in the

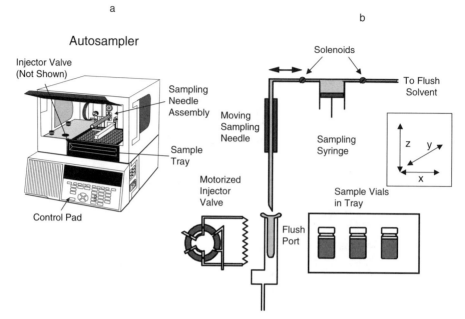

Figure 4.7. A drawing and a schematic of an x-y-z autosampler. Diagram courtesy of PerkinElmer, Inc.

"XY" directions to the samples or the injection valve and up and down in the "Z" direction. It is connected to a sampling syringe driven by a precise stepper motor. The syringe can draw samples or flush solvents depending on the positions of solenoid valves. The injection sequence mimics the operation of a manual injection. First, the needle cleans itself by drawing the flush solvent and empties it into the flush drain port. Then, it moves to the sample vial, withdraws a precise sample aliquot, and delivers it to the injector valve. The motor then turns the valve to inject the sample. Small air gaps are used to segment the sample aliquot from the flush solvent inside the sampling needle. A microprocessor choreographs the precise movements and timings of this injection sequence. This basic design allows rapid sampling from a variety of trays of sample vials or microplates and is central to many modern autosamplers. Another popular design is the integrated-loop autosampler, where a fixed sampling needle forms part of the sampling loop.[2] Sample vials are brought to the sampling needle by a set of robotic arms or a carousel turntable system.

4.3.2 Performance Characteristics and Trends

Autosamplers have made significant progress in reliability and performance. Primary performance criteria are sampling precision and carryover. Precision levels of <0.2% relative standard deviation (RSD) and low carryover of

<0.05% are routinely achievable in most autosamplers. "Carryover" refers to the percentage of the previous sample that is "carried over" to the next sample and must be minimized for some applications (e.g., bioanalytical analysis). Carryover can often be reduced by using a PEEK injector rotor seal. Fast operation (up to 3 injections/min) and sampling from 96-well microplates are desirable for high-throughput screening (HTS) and bioanalytical applications. Many have optional Peltier cooling trays or additional liquid handling capabilities for dilution, standard addition, and derivatization.[2] Some autosamplers for HTS handle a large number of microplates by either using large x-y platforms or robotic plate feeders. Others have multiple sampling probes and injectors to allow parallel injection of four or eight samples simultaneously. Examples are shown in a later section.

4.4 DETECTORS

An HPLC detector measures the concentration (or mass) of eluting analytes[4] by monitoring one of their inherent properties, such as UV absorbance. A detector can be "universal" to all analytes or "specific" to particular classes of analytes. Common detectors and their attributes are listed in the Table 4.2. Early HPLC detectors are spectrometers equipped with small flow cells; however, most modern units are compact and designed solely for HPLC. The ubiquitous UV/visible variable wavelength absorbance and the photodiode array detectors (PDA) are covered in more depth in this section. Note that mass spectrometers (MS) and nuclear magnetic resonance spectrometers (NMR) are discussed in the section on hyphenated systems.

4.5 UV/VIS ABSORBANCE DETECTORS

4.5.1 Operating Principles

The UV/Vis absorbance detector monitors the absorption of UV or visible light in the HPLC eluent. They are the most common detectors since most analytes of interest (e.g., pharmaceuticals) have UV absorbance. A UV/Vis detector consists of a deuterium lamp, a monochromator, and a small flow cell (Figure 4.8a). A monochromator consists of a movable grating or prism that allows the selection of a specific wavelength to pass through the exit slit. A dual-beam optical design is common. Here the light source is split into a sample and a reference beam, and the intensity of each beam is monitored by a separate photodiode. Only the sample beam passes through the sample flow cell. A flow cell (Figure 4.9a) has typical volumes of 2–10 μL and path lengths of 2–10 mm, with quartz lenses serving as cell windows.

The principle for UV/Vis absorption is the Beer's Law,[1] where

Absorbance (A) = molar absorptivity (ε) × pathlength (b) × concentration (c)

Table 4.2. Common HPLC Detectors and Attributes

Detector	Analyte/attributes	Sensitivity
UV/Vis absorbance (UV/Vis)	Specific: Compounds with UV chromophores	ng–pg
Photo diode array (PDA)	Specific: Same as UV/Vis detectors, also provides UV spectra	ng–pg
Fluorescence (Fl)	Very specific: Compounds with native fluorescence or with fluorescent tag	fg–pg
Refractive index (RI)	Universal: polymers, sugars, triglycerides, organic acids, excipients; not compatible with gradient analysis	0.1–10 µg
Evaporative light scattering (ELSD)	Universal: nonvolatile or semivolatile compounds, compatible with gradient analysis	10 ng
Corona-charged aerosol (CAD)	Universal: use nebulizer technology like ELSD and detection of charges induced by a high-voltage corona wire	Low ng
Chemiluminescence nitrogen (CLND)	Specific to N-containing compounds based on pyro-chemiluminescence	<0.1 ng of nitrogen
Electrochemical	Very specific: Electro-active compounds (Redox)	pg
Conductivity	Specific to anions and cations, organic acids, surfactants	ng or ppm–ppb
Radioactivity	Specific, radioactive-labeled compounds	Low levels
Mass spectrometry (MS) MS/MS	Both universal and specific, structural identification; very sensitive and specific	ng–pg pg–fg
Nuclear magnetic resonance (NMR)	Universal, for structure elucidation and confirmation	mg–ng

Absorbance is defined as the negative logarithm of transmittance, which is equal to the ratio of transmitted light intensity and the incident light intensity. Note that absorbance is equal to 1 if 90% of the light is absorbed and to 2 if 99% is absorbed.

Most UV absorption bands correspond to transitions of electrons from $\pi \rightarrow \pi^*$, $n \rightarrow \pi^*$, or $n \rightarrow \sigma^*$ molecular orbitals. Table 4.3 lists some common organic functional groups with chromophoric (light absorbing) properties.

4.5.2 Performance Characteristics

Primary performance characteristic benchmarks of UV/Vis detectors are sensitivity, linear dynamic range, and band dispersion. These are primarily controlled by the design of the optics and the flow cell.

Sensitivity is specified by baseline noise. For years, noise specification has been "benchmarked" at 1.0×10^{-5} absorbance unit (AU) (Figure 4.9b). A wavelength range of 190–600 nm is typical, though sensitivity is substantially lower >400 nm due to a lack of energy of the deuterium source in the visible region. Many detectors allow a secondary tungsten source to increase sensitivity in

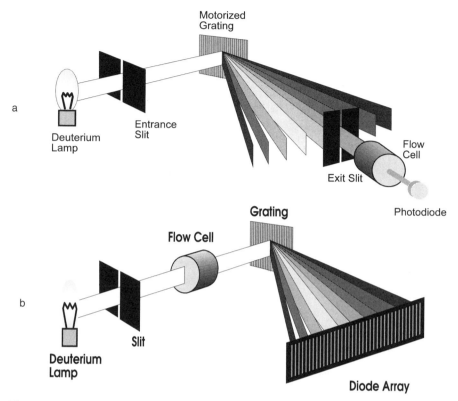

Figure 4.8. A schematic of a UV-Vis absorbance detector (a) and a photodiode array (PDA) detector (b). Diagrams adapted from and reprinted with permission from Academy Savant.

the visible region. Note that when a single wavelength is selected, a typical spectral bandwidth of 5–8 nm actually passes through the flow cell. Increasing the spectral bandwidth by widening the exit slits improves detection sensitivity but reduces linearity or the linear dynamic range (LDR) (Figure 4.9c). Flow cell design is also important for increasing sensitivity since signals are proportional to flow cell path lengths as per Beer's Law. Increasing path lengths with a larger flow cell, however, leads to higher system bandwidth or extra-column band-broadening. Most detectors have optional flow cells for specific applications of semi-micro, microbore, semi-prep, or LC/MS. Drift is defined as the change of baseline absorbance with time and is measured in AU/hour (Figure 4.9b). Drift is typically low in dual-beam detectors ($\sim 1.0 \times 10^{-4}$ AU/hour), which compensates for the long-term change in lamp energy.

4.5.3 Trends in Absorbance Detectors

Sensitivity and linearity performance have improved steadily in recent decades. The benchmark noise level of $\pm 1 \times 10^{-5}$ AU/cm has now been

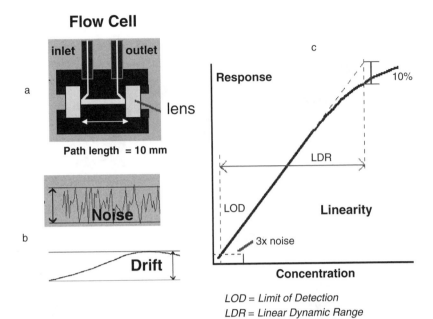

LOD = Limit of Detection
LDR = Linear Dynamic Range

Figure 4.9. (a) Schematic diagram of a flow cell and (b) baseline chromatograms showing noise (magnified) and drift. (c) Chart of UV response with concentration of the analyte injected. Linearity range is from limit of detection (LOD) to the point deviating 10% from the linear response. Diagrams courtesy of Academy Savant.

Table 4.3. Common UV Chromophores

Chromophore	λ_{max}(nm)	Molar absorptivity (ε)
Alkyne	225	160
Carbonyl	280	16
Carboxyl	204	41
Amido	214	41
Azo	339	5
Nitro	280	22
Nitroso	300	100
Nitrate	270	12
Olefin conjugated	217–250	>20,000
Ketone	282	27
Alkylbenzenes	250–260	200–300
Phenol	270	1,450
Aniline	280	1,430
Naphthalene	286	9,300
Styrene	244	12,000

Data extracted from reference 1.

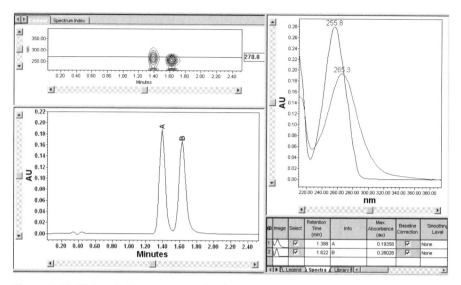

Figure 4.10. Waters EmPower screen showing a contour map, a chromatogram at 270 nm showing the separation of nitrobenzene and propylparaben. Spectra of these two components are shown in the right-hand panel annotated with their respective maximum absorbance wavelengths.

surpassed by many modern detectors. The high end of the linear dynamic range is also extended from a typical level of 1–1.5 AU to more than 2–2.5 AU in some detectors by lowering stray lights and reducing scattering in the flow cell. The typical lifetime of the deuterium lamp is now 1,000–2,000 hours. Many modern detectors have dual- or multiple-wavelength detection capability. Most detectors have features such self-aligned sources and flow cells, leak sensors, and built-in holmium oxide filters for easy wavelength accuracy verification. One significant trend is the design of detectors for parallel analysis by incorporating fiber optics and multiple flow cells (e.g., Waters 2488 detector).

4.6 PHOTODIODE ARRAY DETECTORS

A photodiode array detector (PDA), also known as a diodearray detector (DAD), provides UV spectra of eluting peaks while functioning as a multi-wavelength UV/Vis absorbance detector. It facilitates peak identification and is the prefered detector for method development. Detector sensitivity was lower in earlier models but has improved significantly in recent years.

4.6.1 Operating Principles

Figure 4.8b shows the schematic of a PDA detector where the entire spectrum of the deuterium lamp passes through the flow cell onto a diode array element

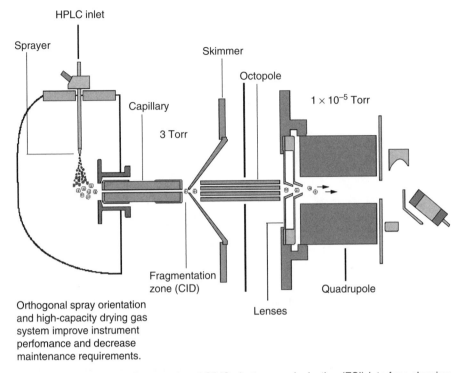

HPLC inlet

Sprayer

Skimmer

Octopole

1×10^{-5} Torr

Capillary

3 Torr

Fragmentation
zone (CID)

Quadrupole

Orthogonal spray orientation
and high-capacity drying gas
system improve instrument
perfomance and decrease
maintenance requirements.

Lenses

Figure 4.11. Schematic diagram of an LC/MS electrospray ionization (ESI) interface showing the nebulization of the LC eluent into droplets, evaporation of the solvent, and the ionization of the analytes, which are pulled inside the MS. Diagram courtesy of Agilent Corporation.

that measures the intensity of light at each wavelength. Most PDAs use a charge-coupled diode array with 512–1,024 diodes (or pixels), capable of spectral resolution of about 1 nm. Spectral evaluation software allows the display of both chromatographic and spectral data in samples (Figure 4.10). These features are integrated into the chromatographic data system and also include automated spectral annotations of λ_{max}, peak matching, library searches, and peak purity evaluation. A summary of features and functions of modern PDAs are:

- Contour maps and 3-D spectral display, for overall evaluation of the entire sample
- λ_{max} annotations, for peak tracking during method development
- Spectral library searches, to aid peak identification
- Peak purity evaluation, by comparing the upslope, apex and downslope spectra, assessment of UV spectral peak purity can give some limited assurance that there is not an impurity with different characteristics co-eluting with the main peak

4.6.2 Trends in PDA Detectors

Modern PDAs have sensitivity performance close to the benchmark level of $\pm 1 \times 10^{-5}$ AU for UV/Vis detectors. Further sensitivity enhancement has stemmed from innovative flow cell design using light-pipe (fiber-optics) technology to extend the pathlength without increasing noise or chromatographic dispersion. By coating the light-pipe with a reflective polymer to allow total internal deflection, very small flow cells can be constructed with long pathlengths (e.g., 50 mm) with excellent dispersive characteristics. One noteworthy development is a programmable slit design, which allows the user to select for either high detector sensitivity (wider slit) or high spectral resolution (narrower slit).

4.7 OTHER DETECTORS

4.7.1 Fluorescence Detector (FLD)

A fluorescence detector monitors the emitted fluorescent light of the HPLC eluent in the flow cell with irradiation of an excitation light at a right angle (see schematic diagram in Figure 4.12a). It is selective and extremely sensitive (picograms to femtograms) but is limited to compounds with strong innate fluorescence.[1,4] A fluorescence detector can be a regular fluorescence spectrophotometer fitted with a small flow cell though most are built specifically for HPLC. A fluorescence detector consists of a xenon source, an excitation monochromator, an emission monochromator, a square flow cell, and a

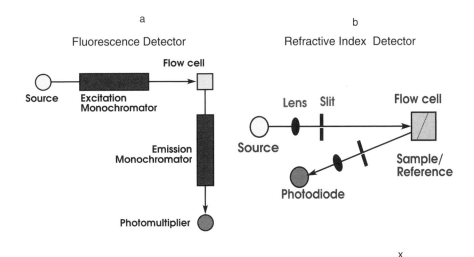

Figure 4.12. (a) Schematic diagram of a fluorescence detector with dual monochromators and a square flow cell. (b) Schematic diagram of a refractive index detector.

photomultiplier for amplifying the emitted photons. The xenon lamp can be a continuous source or a lower wattage pulsed source. The pulsed source is becoming more common because it has more energy in the far UV region and can also measure phosphorescence, chemiluminescence, and bioluminescence. More expensive units have a double monochromator, which can be time-programmed to optimize detection for multiple components. Filters are used instead of monochromators in lower-cost units. Sensitivity is often increased by widening the optical slits (e.g., up to 20 nm in spectral bandwidth).

4.7.2 Refractive Index Detector (RID)

A refractive index detector measures the refractive index change between the sample cell containing the eluting analyte and the reference cell purged with pure eluent (see Figure 4.11b). It has lower sensitivity (0.01–0.1 µg) and is prone to temperature and flow changes.[1,4] It offers universal detection and is used commonly for analytes of low chromophoric activities such as sugars, triglycerides, organic acids, pharmaceutical excipients, and polymers. It is the standard detector in GPC.[15] Modern RID are mostly differential deflection type (Figure 4.11b). Baseline stability has improved significantly in recent years by better thermostating the flow cell. However, its lower sensitivity as compared with other detectors and incompatibility with gradient elution are still the biggest disadvantages.

4.7.3 Evaporative Light Scattering Detector (ELSD)

An evaporative light scattering detector (ELSD) nebulizes the HPLC eluent to eliminate the mobile phase and measures the scattered radiation of a laser beam by the particle stream of all nonvolatile analytes. Its big advantages compared with RI detection are the higher sensitivity (~10 ng) and compatibility with gradient elution. ELSD are used for analytes of low UV absorbance (e.g., sugars, triglycerides), GPC, and high-throughput screening of combinatorial libraries.

4.7.4 Corona-Charged Aerosol Detector (CAD)

A corona-charged aerosol detector (CAD) is a new universal detector introduced by ESA in 2004. It uses a nebulizer technology similar to ELSD, with detection of charges induced by a high-voltage corona wire. Sensitivity is in the low nanogram range. Compared with the ELSD, CAD has the added advantages of higher sensitivity, easier operation (less parameters to optimize), consistent mass response factors to diverse analytes (<±10%), and wider linear dynamic range (4 orders of magnitude). This innovative detector is expected to have significant impacts as a high-sensitivity universal detector and on specific research investigations such as the estimation of response factors of unknown pharmaceutical impurities. Note that CAD, like ELSD, can only detect nonvolatile analytes.

4.7.5   Chemiluminescence Nitrogen Detector (CLND)

Chemiluminescence nitrogen detector (CLND) is a nitrogen-specific detector based on pyro-chemiluminescence technology where N-containing compounds are oxidized to nitric oxide. The nitric oxide then reacts with ozone and emits light of a specific wavelength. It is highly specific and sensitive (<0.1 ng of nitrogen) and yields equimolar response to the nitrogen in the compound. It is particularly useful for the determination of relative response factors of unknown impurities and degradants in pharmaceuticals. Note that CLND cannot be used with any N-containing mobile phase components, such as acetonitrile or ammonium compounds.

4.7.6   Electrochemical Detector (ECD)

An electrochemical detector (ECD) measures the electrical current generated by electroactive analytes in the HPLC eluent between electrodes in the flow cell.[1,4] It offers sensitive detection (picogram levels) for catecholamines, neurotransmitters, reducing sugars, glycoproteins, and compounds with phenolic, hydroxyl, amino, diazo, or nitro functional groups. ECD can be amperometric, pulsed-amperometric, or coulometric type and common electrodes materials are carbon, silver, gold, or platinum, operated in the oxidative or reductive mode. It is capable of high sensitivity and selectivity, though the fouling of electrodes can be a common problem with real samples.

4.7.7   Conductivity Detector

A conductivity detector measures the electrical conductivity of the HPLC eluent stream and is amenable to ppm-ppb levels analysis of ions, organic acids, and surfactants. It is the primary detection mode for ion chromatography.[30]

4.7.8   Radiometric Detector

A radiometric detector, also called a radio-flow detector, is used to measure radioactivity of radioactive analytes in the HPLC eluent passing through a flow cell. Most are based on liquid scintillation technology to detect phosphors caused by the radioactive nuclides. A liquid scintillator can be added post-column with a pump or a permanent solid-state scintillator can be used around the flow cell. This detector is specific only to radioactive compounds and can be extremely sensitive. This detector is used for experiments using tritium or C-14 radiolabeled compounds in toxicological, metabolism, or degradation studies.

[*A Note on Post-Column Reaction Techniques*: A post-column reaction unit is an online derivatization system that supplies reagents to the column eluent into a heated chamber to convert the analytes into more chromophoric forms for higher sensitivity detection. Some common applications of post-column reaction systems are amino acid analysis using ninhydrin (with visible detection), and carbamate pesticide analysis using o-phthaldehyde (with fluorescence detection).]

4.8 HYPHENATED AND SPECIALIZED SYSTEMS

4.8.1 LC/MS, LC/MS/MS

Mass spectrometry with its excellent sensitivity is emerging as one of the most powerful analytical techniques.[16] Its importance was recognized by the awarding of the 2002 Nobel Prize in Chemistry to John B. Fenn and Koichi Tanaka for their research in mass spectrometric methods for biomolecules. The primary difficulties of combining LC and MS have been the interface, and the ionization of analytes in a stream of condensed liquids and transfer of ions into the high vacuum inside the mass spectrometry. Two common LC/MS interfaces are the electrospray ionization (ESI) and the atmospheric pressure chemical ionization (APCI). Figure 4.13a shows a schematic diagram of an

b. Bruker LC/NMR/MS

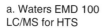
a. Waters EMD 100
LC/MS for HTS

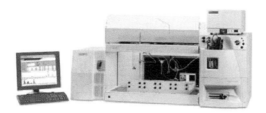

c. Shimadzu Prep LC

d. ABI Vision for Bio-purification

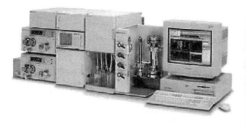

Figure 4.13. Pictures of hyphenated systems and systems for prep LC and bio-purification. Note that the magnet is not shown in the Bruker LC/NMR system.

electrospray ionization interface where HPLC eluent is sprayed diagonally across a capillary inlet of the MS. During the spraying process, solvent molecules are removed with the aid of heat and drying gases while the charged analytes are guided into the mass spectrometer. Common types of MS include the quadrupole, ion trap, triple quadrupole, time of flight (TOF), and Fourier transform MS (FTMS). While most MS have unit mass resolution and have mass range up to ~2,000, TOF and FTMS are capable of higher mass range and much higher mass resolution (1–10 ppm). The pros and cons of each MS type are discussed elsewhere.[16] Figure 4.13a shows an LC/MS system configured for high-throughput screening.

As compared with a single quadrupole LC/MS, the use of LC/MS/MS using a triple quadrupole analyzer can further increase sensitivity and specificity for quantitation of trace analytes in complex matrices. Here the parent ions of the analyte are fragmented, and the daughter ions are monitored for structure determination or quantitative analysis. LC/MS/MS is now the dominant methodology for bioanalytical analysis (drugs and metabolites in physiological fluids and tissues) and offers a generic approach for trace analysis in complex samples. With continued cost reduction and increased reliability, LC/MS/MS will soon have wider impacts in routine testing of pharmaceutical impurities and environmental pollutants.

4.8.2 LC/NMR

The hyphenated technique of LC/NMR is an area of rapid growth due to significant instrument enhancements in recent years.[17] Traditional NMR is used mostly in the off-line mode for relatively pure components and has a minimum sample requirement of several milligrams. Combining LC and NMR extends the technique for online mixture analysis. The primary problem has been the low sensitivity of NMR, which was now extended to microgram and even nanograms levels by probe miniaturization, noise reduction, and other innovations in interface technologies. Many research laboratories now combine HPLC with several detectors such as PDA, MS, and NMR, yielding a powerful problem-solving system (Figure 4.13b).

4.8.3 Other Hyphenated Systems

Several other hyphenated systems are LC coupled with infrared spectroscopy (IR) for compound identification and LC couple with inorganic spectroscopy such as atomic absorption (AA) or inductively coupled plasma (ICP) for studies of metal speciation in samples.

4.8.4 Prep LC and Bio-Purification Systems

The goal of preparative LC is the recovery of purified materials.[6] The quantity of materials for purification (milligram, gram, and kilogram) dictates the size of the column and the type of instrumentation required (semi-prep, prep, and large-scale prep). Prep LC systems (Figure 4.13c) have bigger pumps with large pistons for higher flow rates, injectors with larger sample loops and wider internal channels, detectors with short-pathlength flow cells (to prevent signal saturation), and automatic fraction collectors. While most preparative LC is still conducted manually, some applications are highly automated. For instance, systems designed for the automatic purification of combinatorial libraries may use MS detection with sophisticated software to direct the collection of the target components (see Chapter 7).

Chromatography of proteins often requires the use of high-salt mobile phase, which is corrosive to stainless steel.[7] HPLC systems and columns for this application often use titanium or PEEK to substitute for stainless steel components in their fluidics. Other systems have built-in column switching for isolating active components using affinity/IEC or other tandem two-dimensional chromatography (Figure 4.13d).

4.8.5 Proteomics Systems: Capillary LC and Multi-Dimensional LC

Proteomics research is currently driving HPLC technologies towards high sensitivity, speed, and peak capacities.[8] Capillary LC (sometime termed micro LC or nano LC)[18,19] is used to enhance mass sensitivity needed to analyze minute sample amounts. System requirements for handling extremely low flow rates and small peak volumes mandate the use of specialized HPLC system such as the one shown in Figure 4.14a, which has a micro-pump, a micro x-y-z autosampler, and a PDA detector with a low-volume flow cell. Figure 4.14b illustrates the performance at $12\,\mu L/min$ of this system.

Proteomics, defined as the study of the full set of proteins encoded by a genome, generally involve samples too complex to be sufficiently resolved by a single HPLC column. Currently, the traditional approach is two-dimensional polyacryamide gel electrophoresis (2-D PAGE). An even better approach appears to be multi-dimensional chromatography where the sample is fractionated by the first column and each eluent fraction is subsequently analyzed online by another column in the second dimension. Dramatic resolution increase is realized since the overall peak capacity of the system is the product of the peak capacities of the two dimensions. To maximize resolving power, the two dimensions should be orthogonal (using different chromatographic modes such as IEC and RPC). Since the maximum peak capacity of a high-efficiency column with long gradient time can be as high as 400, a total peak capacity of ~15,000 can be predicted for 2-D system. When coupled with time-of-flight MS and bioinformatics software, these 2-D systems can be a powerful tool for the characterization of very complex protein or peptide mixtures. Figure 4.15a shows an LC/MS system configured for proteomics applications.

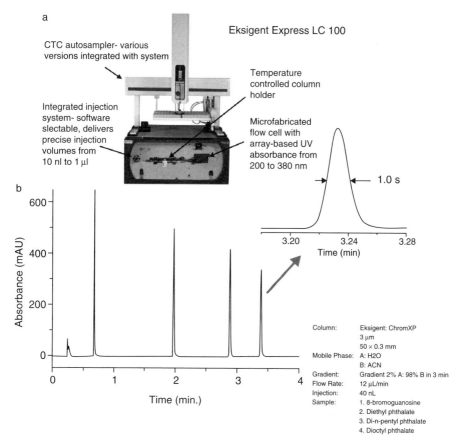

Figure 4.14. A picture of a micro LC system with a chromatogram illustrating the performance of the system. Diagrams courtesy of Eksigent Corp.

4.8.6   High-Throughput Screening (HTS) and Parallel Analysis Systems

High-throughput screening (HTS) in drug discovery is an area of active research pushing HPLC to higher levels of speed and productivity.[20] A typical HTS system consists (Figure 4.13a) of a high-pressure mixing pump, a large platform autosampler for accessing many microplates and multiple detectors (PDA, MS, and ESLD). Typical HPLC columns are short Fast LC narrow-bore columns running full ballistic gradients of 1–5 minutes. Applications of HTS are shown in Chapter 6.

Another generic approach in HTS is "parallel analysis," which enjoyed prominent success in the rapid decoding of the Human Genome by capillary electrophoresis with multiple capillaries in 2000. Figure 4.16a–c shows some available HPLC equipment used for parallel analysis. Systems might include

a. Agilent LC/MS Proteomics Solution

c. Hitachi L-8800 Amino Acid Analyzer

b. Waters Acquity - UPLC

d. Polymer Labs High-Temperature GPC

 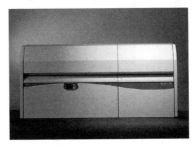

Figure 4.15. Pictures of four specialized applications systems.

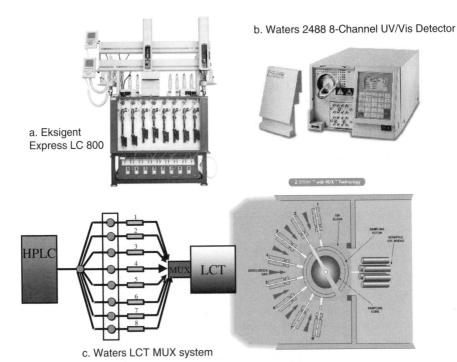

b. Waters 2488 8-Channel UV/Vis Detector

a. Eksigent Express LC 800

c. Waters LCT MUX system

Figure 4.16. (a) Pictures of an Eksigent Parallel LC system and (b) Waters 2488 eight-channel UV/Vis detector. (c) Schematic diagram of a multiplexed parallel LC/MS analysis. Diagrams courtesy of Eksigent and Waters Corporation.

multiple pumps or a multi-channel pumping system, a multi-probe autosampler capable of simultaneous four or eight parallel injections from 96-well microplates, and a multiplexed UV detector and/or MS. The UV detector uses fiber optics technology to accommodate an array of eight flow cells (Figure 4.16b) while the MS multiplexer uses a high-speed valve for sampling each of the eight-column eluent streams (Figure 4.16c).

4.8.7 Ultra-High-Pressure Liquid Chromatography

Ultra-high-pressure liquid chromatography is the most direct way of increasing peak capacity or analysis speed. It was pioneered by the research works of Professors James Jorgenson[21] and Milton Lee. The 6,000-psi pressure limit of the standard HPLC system set the upper limit of peak capacity to ~200–400. Higher system operating pressure allows the use of longer columns for higher resolving power or faster analysis with sub-3μm-particle columns. This approach is embraced by several manufacturers, including a recent introduction of a low-dispersion ultra performance LC (UPLC)™ system shown in Figure 4.15b capable of system pressure up to 15,000 psi.[22] The Waters Acquity system consists of a binary high-pressure mixing pump, a low-dispersion micro-autosampler, and a UV-Vis or PDA detector with a 500-nL lightPipe™ flow cell with a 10-mm pathlength. This UPLC™ system is used with a 1.7-μm particle column with specialized fittings capable of very high-speed LC (sub-minute analysis) or very high resolution (peak capacity <400).

4.8.8 Lab-on-a-Chip

One of the most exciting areas of analytical research is the concept of lab-on-the-chip using micro-fabrication technology.[9] It promises the ultimate in low-cost and high-speed multi-channel analysis. Currently, the most successful applications in commercialized instruments appear to be the analysis of biomolecules using capillary electrophoresis-based chips. HPLC chip-based technology is much anticipated though more technically challenging.

4.8.9 Specialized Applications Systems

A simple gel-permeation chromatography system (GPC) consists of an iso-cratic pump, a GPC column, and a refractive index detector.[15] Specialized software is needed for the calculation of molecular weight averages. For more elaborate analysis using universal calibration or determination of absolute molecular weights, additional detectors such as an online viscometer or a low-angle or multiple-angle light scattering detector can be used.[10] Figure 4.15d shows a high-temperature GPC unit designed for polyolefin analysis at up to 200°C. Additional examples are shown in Chapter 7.

   An automated HPLC dissolution system often uses a transfer module to collect samples from different dissolution vessels into empty HPLC vials preloaded into the HPLC autosampler (e.g., Waters 2695D). The system can also analyze the collected samples while the dissolution experiment is still in progress. These systems are particularly productive for the evaluation of controlled-release dosage forms requiring sampling of multiple time points. Description of dissolution analysis of pharmaceuticals is found in Chapter 6.

Ion chromatography has evolved into a specialized market segment, though HPLC equipment is largely used. In essence, ion chromatography is ion-exchange chromatography with conductivity detection, often with a suppressor for enhancing sensitivity. It is particularly useful in the analysis of low levels of anions and metals in environmental and industrial samples.[11] Applications are shown in Chapter 7.

While most HPLCs are general-purpose instruments, many are dedicated analysis systems packaged with specific columns and reagents. They often come with guaranteed performance from the vendor. Examples are systems for the analysis of amino acids (Figure 4.15c), carbamate pesticides, or sugars (see Chapter 7).

4.9 HPLC ACCESSORIES AND DATA HANDLING SYSTEMS

4.9.1 Solvent Degasser

Dissolved gas in the mobile phase can cause pump malfunctions leading to blending errors and shifting retention times. Degassing by stirring or ultra-sonication under vacuum is inadequate while helium sparging is inconvenient and expensive. Online vacuum degassers are now built-in accessories in many integration HPLC systems. In these vacuum degassers, solvents passing through tubes of semiporous polymer membranes inside a chamber, which is evacuated to eliminate dissolved gaseous molecules. Their reliability has improved and degasser tube volumes have been reduced to <1 mL. Note that helium degassing and pressurization systems are still required for pumping very volatile solvents (e.g., pentane, methylene chloride), or labile reagents (e.g., ninhydrin) and in some trace analysis gradient applications.

4.9.2 Column Oven

A column oven is required for most automated assays to improve retention time precision. Column temperatures of 30–50°C are typical. Temperatures higher than 60°C are often used to increase flow and column efficiency. Sub-ambient operation is used in chiral separations to enhance selectivity. Column ovens operate either by circulating heated air or direct contact (clam-shell type). Solvent preheating is achieved by passing the mobile phase through a long coiled tube embedded onto the heating element before the column. New trends are toward wider temperature ranges (e.g., 4–100°C) using Peltier devices.

4.9.3   Column Selector Valve

Column selector valves can be added on as accessories to allow column switching for multi-dimensional chromatography or for automatic column selection to facilitate methods development. Column switching valves should be located inside the column oven if possible.

4.9.4   Data Handling and HPLC Controllers

Figure 4.17 shows the historical development of chromatographic data-handling devices ranging from a strip chart recorder, an electronic integrator, and a PC-based workstation to a client-server network system.[23] This has also been a progression of increasing sophistication and automation. While the chart recorder requires manual measurement of peak heights, the integrator has built-in algorithms for peak integration, calculation, and report generation. A PC-based workstation often incorporates additional functions of data archiving and full HPLC system control with a user-friendly graphic interface

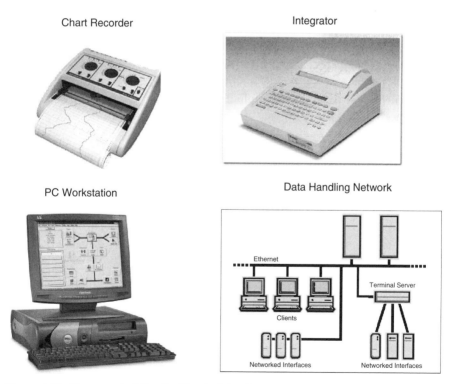

Figure 4.17. Diagrams depicting the historical development of chromatography data handling systems.

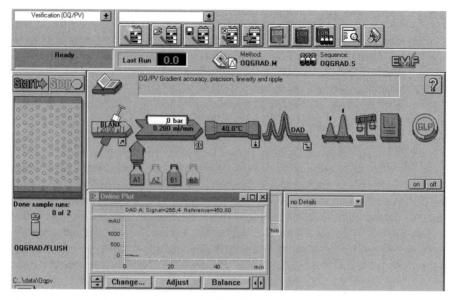

Figure 4.18. Graphical user interface of Agilent ChemStation showing the status of various HPLC modules in the system and a real-time chromatographic plot.

(Figure 4.18). For large laboratories with many chromatography systems, a centralized data network system is becoming the preferred solution to ensure data security and regulatory compliance. The reader is referred to reference 23 for more details on the topic.

4.10   INSTRUMENTAL BANDWIDTH (IBW)

The instrumental bandwidth (IBW) or system dispersion has become important in recent years with the popularity of shorter and narrower columns. While band-broadening within the column is innate to the chromatographic process, broadening that occurs outside the column (extracolumn band-broadening) is highly undesirable as it reduces the separation efficiency of the column.[16] Extracolumn band-broadening is caused by the parabolic flow profiles in connection tubing, injectors, and detector flow cells. Its effect can be minimized by using small-diameter connection tubing, a well-designed injector, and a small detector flow cell. Figure 4.19a shows the effect of extracolumn band-broadening on the Fast LC separation of four parabens. An optimized HPLC system with a semi-micro 2.4-µL flow cell yields sharp peaks and good resolution while the same instrument equipped with an 8-µL standard flow cell shows considerably broader peaks.

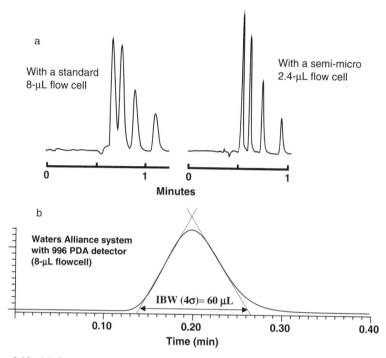

Figure 4.19. (a) Comparative chromatograms showing the deleterious effect of instrumental bandwidth to the performance of a Fast LC separation of four paraben antimicrobials. Reprint with permission from reference 24. (b) The instrumental bandwidth of a Waters Alliance HPLC system with a 966 PDA detector with a standard 8-µL flow cell.

This effect of system bandwidth can be calculated using the additive relationship of variances where the total variance of the observed peak is equal to the sum of the true column peak variances plus the variance of the IBW:

$$\sigma^2_{total} = \sigma^2_{column} + \sigma^2_{IBW}$$

Here, the variance of the IBW is the summation of variances from the injector, the detector, and the connection tubing:

$$\sigma^2_{IBW} = \sigma^2_{injector} + \sigma^2_{detector} + \sigma^2_{connection\ tubing}$$

This instrumental bandwidth can be physically measured as follows:

- Replace the column with a zero dead-volume union
- Set the HPLC system to 0.5 mL/min, UV detection at 254 nm, and a data rate of 10 points per second

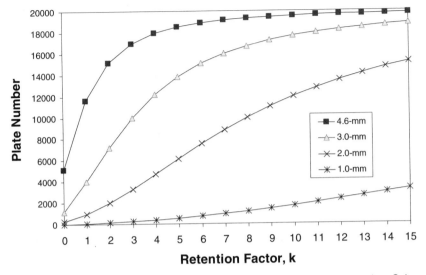

Figure 4.20. Chart showing the effect of system dispersion on column efficiencies. Columns are 150 mm long and vary from 1.0 to 4.6 mm inner diameters and retention factors k of 1–15. Columns are packed with 3-μm particles. The HPLC used has an IBW (4σ) of 60 μL.

- Inject a 1-μL aliquot of a 0.5% caffeine or uracil solution and record the resulting trace

Calculate the IBW (4σ) using the tangent method as shown in Figure 4.19b. An IBW of ~60 μL was found for this HPLC. Figure 4.20 shows a chart plotting column efficiencies of 150-mm long 3-μm columns of various inner diameters versus retention factors (k) on an HPLC system with a 60-μL IBW. Note that efficiency loss from instrumental dispersion can be severe for small inner diameter columns where peak volumes are smaller. Note that peaks of low k are affected more by system dispersion since peak volumes are proportional to k.

The IBW of most HPLC systems can be reduced by these modifications:

- Replace connection tubing with shorter lengths of 0.005″–0.007″ i.d. tubing
- Replace the detector flow cell with an optional semi-micro flow cell (2–5 μL)
- Reduce the sample injector volumes <20 μL for isocratic analysis

Further reduction of IBW might involve the replacement of the existing injector with a low-dispersion micro-injector or an autosampler designed for micro LC.

In gradient analysis, the sample band is typically reconcentrated at the head of the column, so the dispersive contribution of the injector and tubing before

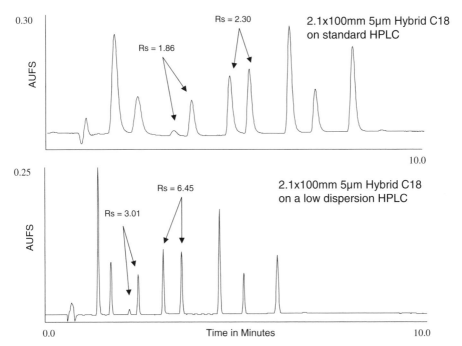

Figure 4.21. Effect of instrumental dispersion on gradient analysis on a short 2-mm column. (a) Analysis on a "standard" HPLC system. (b) Analysis of the same sample on a low-dispersion system (Waters Acquity). Diagram courtesy of Waters Corporation.

the column is not important as only the dispersion of the detector flow cell and its connection tube to the column outlet is significant. Still, for small columns of inner diameters $<2\,mm$, the peak resolution performance discrepancy can be dramatic, as shown in Figure 4.21.

4.11 TRENDS IN HPLC EQUIPMENT

Table 4.4 summarizes some of the trends of HPLC equipment. HPLC is a mature analytical technique based on well-developed instrumentation with a definitive trend toward higher performance, better reliability, and easier maintenance. Demands in life sciences and drug discovery are driving HPLC technologies toward higher resolution, sensitivity, and throughput. Recent innovations in micro LC, ultra-high-pressure chromatography, multidimensional chromatography, parallel analysis, and laboratory-on-a-chip are direct responses to these demands. Research laboratories are using more information-rich detectors such as photodiode array, mass spectrometer, and NMR to generate more data per sample run. The impacts of LC/MS and LC/MS/MS are particularly pivotal in revolutionizing the research processes in drug dis-

Table 4.4. Trends in HPLC and Data Handling Equipment

Higher Performance Modules • Ultra-high-pressure pumps, more precise and faster autosamplers, higher sensitivity detectors • Miniaturization and simplification • Easy validation/maintenance	Vendor Support • Complete validation documentation, IQ/OQ/PQ support • Prompt technical and service support
Automated and Ultra-performance System • Single-point control and graphical user interface • Parallel, ultra-high-pressure, micro and nano-LC. Multi-dimensional, Lab-on-a-Chip • Automated method development	Technology Drivers • Drug discovery, combinatorial, high-throughput screening, proteomics • Bioanalytical analysis • LIMS and paperless laboratory
Information-Rich Detectors • Diode array, MS, MS/MS, NMR, ELSD • Multi-wavelength UV/Vis, PDA • Use of multiple detection in each run	Sophisticated Data handling • Networking with centralized archiving and data retrieval • Accessible from multiple locations and the Internet • Sophisticated integration algorithm, relational database, 21CFR Part 11 compliance, custom reporting

covery, bioanalytical analysis and proteomics. With the increased reliability and lower pricing, LC/MS is rapidly becoming a standard routine analytical tool. Client/server data networks are installed for better data security, regulatory compliance, and easier accessibility from multiple locations.

4.12 MANUFACTURERS AND EQUIPMENT SELECTION

Major manufacturers of HPLC instruments include Waters, Agilent (formerly Hewlett Packard), and Shimadzu, PerkinElmer, Thermo, Beckman, Varian, Hitachi, Jasco, Dionex, Gilson, Scientific Systems (SSI), and Isco. The Internet addresses of these companies can be found in the reference section. HPLC is a mature technology and most manufacturers have highly reliable products with sufficient performance and feature sets to be competitive in the market place. However, there can still be significant differences between the vendors on these performance characteristics on systems (dwell volume, dispersion), pumps (low flow, seal life), autosamplers (carryover, speed, sample capacity, minimum sample volume), and detectors (sensitivity, gradient baseline shift).

Equipment selection is often based on performance specifications, pricing, features, or vendor's technical or service support. Major companies tend to purchase HPLC from one or two vendors to reduce cost in system qualification, equipment service, and operator training. The compatibility to an existing chromatographic data-handling network often dominates the purchase decision for HPLC equipment.

4.13 SUMMARY

This chapter provides an overview of modern HPLC equipment, including the operating principles and trends of pumps, injectors, detectors, data systems, and specialized applications systems. System dwell volume and instrumental bandwidth are discussed, with their impacts on shorter and smaller diameter column applications. The most important performance characteristics are flow precision and compositional accuracy for the pump, sampling precision and carryover for the autosampler, and sensitivity for the detector. Manufacturers and selection criteria for HPLC equipment are reviewed.

4.14 REFERENCES

1. D.A. Skoog, F.J. Holler, and T.A. Nieman, *Principles of Instrumental Analysis, 5th Edition*, Harcourt College Publishers, Fort Worth, TX, 1997.
2. M.W. Dong, in S. Ahuja and M.W. Dong, eds., *Handbook of Pharmaceutical Analysis by HPLC*, Elsevier, Amsterdam, 2005, Chapter 4.
3. J.M. Miller and J.B. Crowther, *Analytical Chemistry in a GMP Environment: A Practical Guide*, Jossey-Bass, New York, 2000.
4. R.P.W. Scott, *Chromatographic Detectors. Design, Function, and Operation*, Marcel Dekker, New York, 1996.
5. M.S. Lee, *LC/MS Applications in Drug Development*, Wiley-Interscience, Hoboken, NJ, 2002.
6. G. Guiochon, S. Goshen, and A. Katti, *Fundamentals of Preparative and Non-linear Chromatography*, Academic Press, Boston, 1994.
7. M.A. Vijayalakshmi, *Biochromatography: Theory and Practice*, Taylor & Francis, London, 2002.
8. J.-C. Sanchez, G.L. Corthais, and D.F. Hochstrasser, *Biomedical Applications of Proteomics*, John Wiley & Sons, Hoboken, NJ, 2002.
9. E. Oosterbroek and A. Van Den Berg, *Lab-on-a-Chip: Miniaturized Systems for (Bio) Chemical Analysis and Synthesis*, Elsevier Science, Amsterdam, 2003.
10. S. Mori and H.G. Barth, *Size Exclusion Chromatography*, Springer-Verlag, Berlin, Germany, 1999.
11. J. Weiss, *Ion Chromatography, 2nd Edition*, VCH, Weinheim, Germany, 1995.
12. W.R. LaCourse, *Anal. Chem.*, **74**, 2813 (2002).

13. *Market Analysis and Perspectives, Laboratory Analysis Instrument Industry, 1998–2002, 5th Edition*, Strategic Directions International, Inc., Los Angeles, CA, 1998.

14. HPLC equipment, CLC-30, (Computer-based Instruction), Academy Savant, Fullerton, CA.

15. W.M. Reuter, M.W. Dong, and J. McConville, *Amer. Lab.* **23(5)**, 45 (1991).

16. L. Zhou, in S. Ahuja and M.W. Dong, eds., *Handbook of Pharmaceutical Analysis by HPLC*, Elsevier, Amsterdam, 2005, Chapter 19.

17. M. Wann, in S. Ahuja and M.W. Dong, eds., *Handbook of Pharmaceutical Analysis by HPLC*, Elsevier, Amsterdam, 2005, Chapter 20.

18. R. Grimes, M. Serwe, and J.P. Chervet, *LC.GC* **15(10)**, 960 (1997).

19. Waters CapLC System, Waters Corp., Milford, MA, April 2000, 72000133en.

20. R. Kong, in S. Ahuja and M.W. Dong, eds., *Handbook of Pharmaceutical Analysis by HPLC*, Elsevier, Amsterdam, 2005, Chapter 17.

21. J. E., Jorgenson, *Anal. Chem.*, **69**, 983 (1997).

22. Waters Ultra Performance Liquid Chromatography Acquity System, Waters Corp., Milford, MA, 2004.

23. R. Mazzarese, in S. Ahuja and M.W. Dong, eds., *Handbook of Pharmaceutical Analysis by HPLC*, Elsevier, Amsterdam, 2005, Chapter 21.

24. M.W. Dong, *Today's Chemist at Work*, **9(2)**, 46 (2000).

4.15   INTERNET RESOURCES

http://www.waters.com
http://www.chem.agilent.com
http://www1.shimadzu.com/products/lab/hplc.html
http://instruments.perkinelmer.com/ai/chrom/index.asp
http://www.thermo.com
http://www.beckman.com
http://www.varianinc.com
http://www.hitachi-hhta.com
http://www.jascoinc.com
http://www.dionex.com
http://www.gilson.com
http://www.esainc.com/
http://www.rheodyne.com
http://www.isco.com
http://www.sedere.com/
http://www.polymerlabs.com/
http://www.alltechweb.com
http://www.microlc.com
http://www.pickeringlabs.com/

5

HPLC OPERATION GUIDE

Modern HPLC for Practicing Scientists, by Michael W. Dong
Copyright © 2006 John Wiley & Sons, Inc.

5.1 SCOPE

This chapter describes procedures for mobile phase preparation and operation of common HPLC modules as exemplified by best practices of experienced practitioners. Concepts in qualitative and quantitative analysis are described together with trends in chromatographic data analysis and report generation. Environmental and safety concerns are summarized. Guidelines for increasing HPLC precision and avoiding pitfalls in trace analysis are described. The goal is to offer the laboratory analyst a concise operating guide to a higher degree of success in HPLC analysis. The reader is referred to detailed discussions on this topic in reference books,[1–6] manufacturers' operating manuals, training video/software,[7–8] training courses and trade journals.[9]

5.2 SAFETY AND ENVIRONMENTAL CONCERNS

5.2.1 Safety Concerns

The general safety concerns in the HPLC Lab are in-line with most analytical laboratories dealing with small sample sizes.[10] The high-pressure operation of the HPLC instrument usually does not pose a significant safety risk since small volumes of liquids are used and the units are designed for these conditions. Typical safety risks involve external conditions such as high electric voltages in close proximity with the mobile phase, which can be a flammable or combustible liquid. The operation and set-up of HPLC modules should be in compliance with local, state, and national fire codes such as NFPA 30, NFPA 45, and/or NFPA 70 (National Electrical Codes).

The toxicity of organic solvents used in the mobile phase needs to be reviewed and integrated into safe laboratory practices and procedures. The Material Safety Data Sheets (MSDS) of each solvent should be consulted prior to use, preferably with a safety and health professional who is knowledgeable about workplace chemical exposures and fire safety. The two common RPLC solvents, acetonitrile and methanol, can be used safely in the lab if good engineering control and appropriate personal protective equipment (PPE) are being employed. Other solvents used in NPC or GPC like tetrahydrofuran (THF), methylene chloride, dimethyl sulfoxide (DMSO), and dimethyl formamide (DMF) need to be handled with a higher level of work place safety and environmental control. In the case of methylene chloride, OSHA (Occupational Safety and Health Administration) has a specific set of regulations for workplace monitoring and control.

As a general rule, the following practices should be considered when handling mobile phase solvents:

• Wear safety glasses and gloves when handling toxic or corrosive chemicals

a

b

Carbon
cartridge
evaporative
control

Waste tag

Solvent inlet

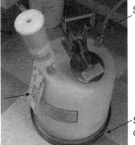

Secondary
containment

Figure 5.1. (a) Solvent cycletainer; (b) Justrite solvent disposal can with secondary containment. Bottom photo in (b) courtesy of Joe Grills of Purdue Pharma.

- Select personal protective equipment (PPE) that is compatible with the solvents being used
- Use appropriate respiratory protection when handling substances with acute toxicities
- Volatile, flammable, and toxic organic solvents should be handled in a laboratory fume hood and/or other systems designed specifically for these applications

Most HPLC-grade solvents are available in 1-gallon glass bottles and should be transported in an appropriate secondary container such as a rubberized carrier. Many common solvents are available in stainless steel "kegs" or cycletainers (Figure 5.1a), which are pressurized with nitrogen for convenient dispensing in the laboratory. Note that NFPA 45 requires that any dispensing of flammable liquid in a laboratory must use local exhaust ventilation, such as a laboratory fume hood. A common cycletainer size is 19-L or 5-gallon. These should be transported with the use of safety carts.

The potential danger for flask implosion during vacuum filtration and degassing should be noted. This danger can be substantial when vacuum filtering larger volumes (i.e., 4 L) or if non-vacuum grade glassware is mistakenly used. Coated glass equipment and additional safety shielding needs to be considered for this operation.

5.2.2 Environmental Concerns

The storage and disposal of all chemicals and solvent wastes must be in accordance with applicable codes and regulations such as RCRA (Resource Conservation and Recovery Act).[11] Waste flammable solvents should be stored in rated flammable liquid containers while they are being filled during the HPLC operation. These containers should be properly labeled and transported to a hazardous waste accumulation area within three (3) days after they are filled. Figure 5.1b shows an example of how a solvent disposal safety can be equipped with an evaporative control and a secondary containment system (a bottom tray). Acids and bases should be stored in specially designed anticorrosive cabinets and segregated to prevent inadvertent mixing from spills. The use of a mobile phase recycler should be explored for less demanding isocratic analyses. The use of smaller HPLC columns should be encouraged whenever possible to conserve solvent usage and to minimize hazardous waste generation.

5.3 MOBILE PHASE PREPARATION

5.3.1 Mobile Phase Premixing

Premix mobile phase by measuring the volume of each solvent separately and combining them in the solvent reservoir.[12] This is particularly important when mixing organic solvents with water because of the negative ΔV of mixing. For example, prepare 1 L of methanol/water (50:50) by measuring 500 mL of methanol and 500 mL of water separately in a measuring cylinder and combine them together. Do not pour 500 mL of methanol into a 1-L volumetric flask and fill it to volume with water as more than 500 mL of water is needed due to the shrinkage of solvents upon mixing.

5.3.2 Buffers

The use of buffers in the mobile phase is required for any samples containing acidic or basic analytes (see Chapter 2). Table 2.3 summarizes the common buffers with their respective pK_a, UV cutoffs, and compatibility with mass spectrometer (MS).

Since a buffering range is ± 1–1.5 pH unit of its pK_a, picking the right buffer for the desired pH is important. A concentration of 10–20 mM is sufficient for most applications.[2,5] Any pH adjustments should be made in the aqueous buffer alone, before mixing it with any organic solvents. Beware of the possibility for buffer precipitation when mixing it with organic solvents such as acetonitrile (ACN)—the upper limit should be 70–80% ACN. Caution particularly should be exercised in blending ACN with phosphate buffers. Limit the concentration of the phosphate buffer to <15 mM to prevent precipitation, which can cause check valve malfunctions. Buffered mobile phases typically last 3 to 10 days pending absorption of carbon dioxide or potentials for bacterial growth.

5.3.3 Filtration

Filtration through a 0.45-μm membrane filter of all aqueous mobile phases containing buffers or ion-pairing reagents is highly recommended. Use cellulose acetate filters for aqueous solvents, and either PTFE or nylon filters for organic solvents. Filtration of any HPLC-grade solvents is not recommended since they are typically prefiltered by the manufacturers. Use all-glass filtration apparatus and 47-mm or 95-mm membranes with thick-wall vacuum-grade filtration flasks. Examine the flask for chips or cracks before use to prevent chances of implosion. Figure 5.2 shows a typical all-glass filtration apparatus and an alternate vacuum system from Kontes, which also serves as a solvent reservoir.

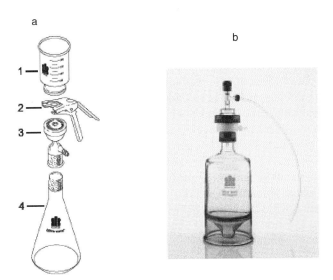

Figure 5.2. (a) All-glass solvent filtration flask using 0.45-μm membrane filters. (Legends: 1. funnel, 2. clamp, 3. filter holder, 4. vacuum flask). (b) Kontes microfiltration system, which can also be used as a mobile phase reservoir. Mobile phase can be pulled into the flask from the Teflon tube without having to pour in from the top in traditional vacuum filtration flask.

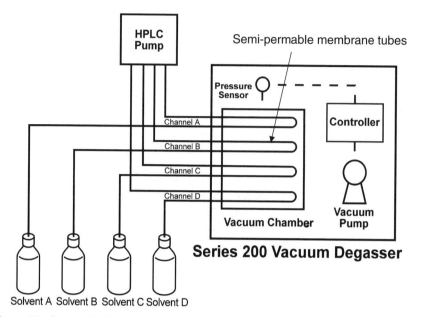

Figure 5.3. Schematic diagram of an online vacuum degasser. Older systems have relatively large semipermeable vacuum tubes of >10 mL. New systems have smaller tubes <1 mL. Note that for effective degassing, all four lines should be filled with solvents. Diagram courtesy of PerkinElmer, Inc.

5.3.4 Degassing

Mobile phase degassing is critical for accurate pump blending and gradient operation.[12] The most convenient way for solvent degassing is by using an online vacuum degasser (see schematic diagram in Figure 5.3). Note that the vacuum membrane tube volumes in the degasser should be purged when changing solvents and most vacuum degasser can be left "ON" all the time. Helium sparging and pressurization is also very effective and might be the only viable mean for pumping volatile solvents (e.g., pentane or methylene chloride) or highly labile reagents (e.g., ninhydrin reagents). Degassing by vacuum filtration or sonication/stirring under vacuum is only partially effective and adequate for isocratic operation. Solvents after this type of degassing will "re-gas" within a few hours and might cause blending error, particularly in low-pressure mixing pumps.

5.4 BEST PRACTICES IN HPLC SYSTEM OPERATION

This section summarizes best practices and standard procedures used by experienced practitioners in HPLC system operation. The information is presented as check lists categorized by the module or the column. An additional section summarizes guidelines for enhancing precision of retention time and peak areas.

5.4.1 Pump Operation

- Place solvent line sinkers (10-µm filters) into intermediary solvent reservoirs when switching solvents to prevent cross-contamination or buffer precipitation.
- Cap the solvent reservoirs or at least cover with Parafilm to minimize atmospheric contamination and evaporation.
- Turn on the online vacuum degasser.
- Set upper pressure limit (typically at 3,500–4,000 psi)
- Set lower pressure limit (typically at 100 psi to trigger pump shutdown if solvent runs out).
- Perform "dry" prime by opening the prime/purge valve and draw out ~10 mL from each solvent line into a syringe (needed only if solvent lines are dry).
- Perform wet prime to purge line with the new solvents when changing the solvent reservoirs.
- Program the pump to purge out the column and shut down the pump after the sample sequence is complete.

Cautions when using buffers:

- Flush buffers from column and system with water.
- Do not let buffers sit in the HPLC system due to the danger of precipitation.
- Always turn on piston seal wash feature to prolong seal life (if available).

5.4.2 HPLC Column Use, Connection, and Maintenance

The following operating guides are recommended for maintaining reversed-phase columns. Consult other references[2,3] or the vendor's column instructions for other column types and for special precautions or column regeneration guides.

5.4.2.1 Column Use

- Store reversed-phase columns in acetonitrile or methanol or in a mixture of water and organic solvents.
- Cap unused columns with "closed" fittings to prevent columns from drying out.
- Always flush the column with a strong solvent (methanol) before use to eliminate any highly retained analyte.

- Use guard column or in-line filter if "dirty" samples are injected. The guard column should preferably be packed with the same packing as the analytical column (see Chapter 3).

5.4.2.2 Column Precautions
- Do not exceed the pH range of the column (typically pH of 2.5–8 for silica-based columns).
- Do not exceed the temperature limits of most columns, usually 60–80°C.
- Never let buffers sit immobile inside columns.
- Only change column temperatures when pressurized.
- *Note:* Many modern HPLC silica-based columns can be used at a wider pH range of 1.5–10. See Figure 3.11).

5.4.2.3 Column Connection
- Use 1/16″ compression nuts and ferrules from Parker, Swagelok, or Rheodyne. These are compatible and interchangeable with each other. Use Waters fittings for Waters columns.
- Use short lengths of 1/16″ o.d. and 0.010″ i.d. Stainless or PEEK tubing for general applications and 0.004″–0.007″ tubing for narrowbore and Fast LC columns.
- Stainless steel tubing should be square-cut with a cutting wheel and have the cut end "deburred" or polished. It is difficult to cut steel tubing with i.d. <0.010″, however, precut tubing can be purchased.
- Make sure that the tubing "bottoms out" inside the fitting when attaching a new ferrule. See Figure 5.4 for ferrule, shapes and correct seating depths of the various fittings.
- Waters and Valco fittings have different ferrule shape and seating depths and are not interchangeable with other "standard" fittings (Figure 5.5a).
- Plastic finger-tight fittings (e.g., UpChurch, Figure 5.5b) are convenient and easy to use. They are excellent if columns from difference vendors are used. They do have a tendency to slip at higher pressure.
- PEEK tubing is convenient and flexible but has a lower pressure rating. PEEK is not compatible with DMSO, THF, and methylene chloride.

5.4.2.4 Column Maintenance and Regeneration
- Typical column lifetime is 3–24 months or 1,000–3,000 injections, depending the type of mobile phase used and samples injected.
- Column performance (efficiency) decreases with time as signified by increased back-pressure and peak widths.
- Monitor the column back-pressure and efficiency performance and take corrective action immediately; don't wait until the column is plugged. If pressure is abnormally high, try back flushing the column as soon as

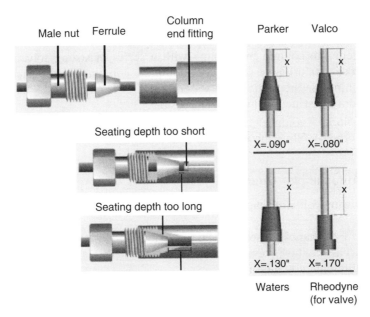

Figure 5.4. Schematic diagrams of column connection hardware and the importance of having the correct seating depths of the ferrule. If the seating depth is too low, additional void volume is created. If the seating depth is too high, leaks might occur. Note that various manufacturers offer fittings of different seating depths and ferrule shapes, which might not be interchangeable. Diagram reprinted from UpChurch.com.

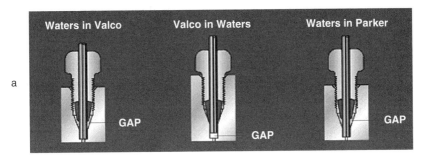

Figure 5.5. (a) Diagram demonstrating problems associated with using HPLC fittings from manufacturers with incompatible formats. Reprinted with permission from Savant Academy. (b) Diagram of examples of finger-tight HPLC sittings. Diagram courtesy of Upchurch Scientific.

possible with a strong solvent. In some columns, the inlet frits can be replaced if plugged.

- Regenerate column by flushing with a series of strong solvents might restore performance of a contaminated column. For reversed-phase columns, use the sequence of water, MeOH, CH_2Cl_2, and MeOH. Consult vendor's instructions whenever possible.
- If column "voiding" occurs in the inlet, some performance might be restored by filling the inlet void with a similar packing.

5.4.3 Autosampler Operation

The following guidelines are recommended for autosamplers[12]:

- Use a methanol or acetonitrile and water mixture without buffer (e.g., 50%) as the flush solvent for most autosamplers. Degas the flush solvent to eliminate any bubbles.
- Fill each vial with enough sample solution for all injections.
- Purge the injector daily and before sample analysis to remove bubbles in sampling syringe. This is critical for sampling precision.
- Typical injection volume range is 5–50 µL.

Cautions:

- Some autosamplers use a side port needle that might require at least a 0.5-mL sample volume in a 2-mL vial (e.g., Waters Alliance).
- If excessive carryover is encountered, explore the use of pre- or post-injection flush or an alternate rotor seal (e.g., PEEK vs. Vespel).
- Injection sizes of <5 µL might lead to poorer precision (see Figure 5.6).
- For large volume injection, a larger sampling syringe and/or sampling loop might be required.
- Put a warning label outside the autosampler if a nonstandard sampling syringe or loop is installed.

5.4.4 Detector Operation

The following general guidelines are suggested for UV/Vis absorbance or photodiode array (PDA) detectors. Consult vendor's manuals for other detector types.

1. Turn lamp on for at least 15 min to warm up before analysis.
2. Set to the appropriate wavelength and detector response time (i.e., 1–2 sec). Use response time of 0.1 or 0.5 sec for Fast LC and data rate of >5 points/second.

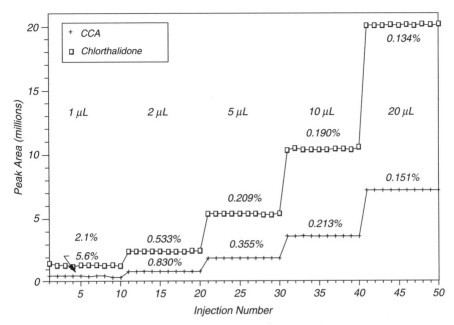

Figure 5.6. Diagram illustrating typical autosampler precision vs. injection volumes. CCA, 4'-chloro-3'-sulfamoyl-2-benzophenone carboxylic acid, is a hydrolysis product for chlorthalidone. Note that peak area precision increases (lower RSD) with larger injection volumes. This was caused by the finite sampling volume precision of the autosampler sampling syringe and stepper motor, which was about 0.01 μL for this particular autosampler. Reprinted with permission from reference 12. See the same reference for additional experimental details.

3. Set wavelength range of PDA (e.g., 200–400 nm) for general method development. Set spectral resolution of 1 nm if spectral data are collected. Use wider spectral resolution (e.g., 5 nm by bunching the signals from adjacent pixels) for better chromatographic sensitivity.

4. Lamps typically last 12 months or over 1,000 hours. Replace lamp if sensitivity loss is observed. Aged lamps typically yield higher baseline noise.

5. Shut off lamps when not in use to increase lifetime.

6. Use an optional semi-micro flow cell for Fast LC or micro LC. Put label on detector front panel signifying the use of a nonstandard flow cell and pathlength.

5.4.5   System Shutdown

The following procedures are recommended for system shutdown:

- Turn off column oven first and let the column cool down before turning off the pump flow.
- Flush the buffer out of the system and stop the flow.

- Keep idle column in methanol or methanol/water.
- Turn off the lamp to preserve source life.
- Leave degasser turned on if possible.

5.4.6   Guidelines for Increasing HPLC Precision

Guidelines for increasing HPLC precision for retention time and peak area are presented below. The reader is referred to references 12 and 13 for details on the theories and case studies.

5.4.6.1   Guidelines for Improving Retention Time Precision
- Use a precise pump (flow precision <0.3%) and a column oven (±0.1°C).[12,13]
- The use of a column oven is mandatory for automated unattended analysis and particularly if laboratory is not temperature controlled.
- Most column ovens are operative from ambient +5°C upward. Minimum specified temperature should be at least 30°C. Exceptions are Peltier ovens with cooling capability.
- Use premixed mobile phases for isocratic analysis to reduce mobile phase variations.
- For pump blending or gradient applications, thorough mobile phase degassing is critical.
- The column can cause fluctuations of retention time if it is failing, not fully equilibrated or contaminated.
- For gradient RPC analysis, at least 3–5 column void volumes of the initial mobile phase must be passed to re-equilabrate the column.
- For gradient applications at low flow rates (<0.1–0.2 mL/min), piston size, and mixing volumes of the LC pumps can be the limiting factors. Use specialized pumps for microbore and capillary columns (see Chapter 4).

5.4.6.2   Guidelines for Improving Peak Area Precision
- Use a precise autosampler (<0.5% RSD) and injection volumes >5 μL (Figure 5.6).[12,13] Note that larger volume injection is more precise since the volume sampling mechanism of the autosampler tends to be the limiting factor (e.g., if the resolution of the sampling syringe is ±0.01 μL, then the best precision obtainable from a 1-μL injection is ±1% or ±0.1% for a 10-μL injection).
- Degas flush solvents used for the autosampler.
- Purge the sampling syringe to eliminate any air bubbles before sample analysis.
- Avoid using airtight sample vials with silicone septa to avoid the possibility of creating a partial vacuum during sample withdrawals.

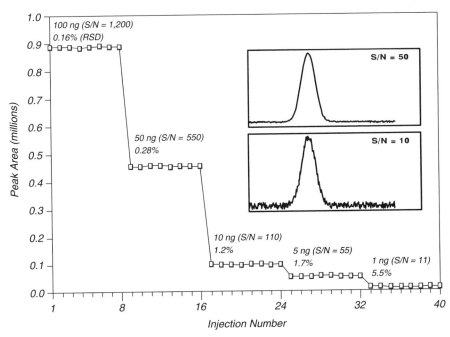

Figure 5.7. Diagram illustrating typical autosampler precision vs. peak signal/noise ratio (keeping injection volume constant at 10 μL). Note that the peak area precision worsens (increasing RSD) because precision was limited by the statistical variation of integration of noise peak when the signal-to-noise ratios (S/N) are less than 100. Reprinted with permission from reference 12. See the same reference for additional experimental details.

- Use vials with thin septa to prevent sample evaporation particularly if the sample solvent is volatile. Avoid sampling the same standard vial excessively. Use multiple standard vials for long sample sequences.
- Improve signal/noise ratio of peaks to >50 if possible and use a data sampling rate of >8 points/peak (Figure 5.7).
- Plot baseline with chromatograms in reports and watch out for integration problems. Use more advanced integration algorithms (e.g., ApexTrack) if possible for complex chromatograms with sloping baselines.

5.5 FROM CHROMATOGRAMS TO REPORTS

While the first goal of most analysis is to obtain a chromatogram with adequate separation, much work remains to convert the chromatographic raw data into useful quantitative information, mostly as chromatography reports. In the past two decades, both HPLC instrumentation and chromatographic data systems have made tremendous improvements in performance and refinements. Chapter 4 gives a brief history of its evolutionary progress of data

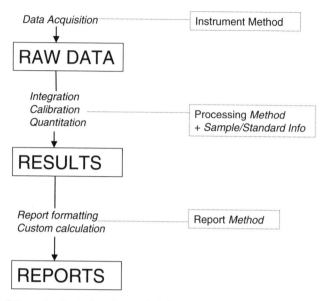

Figure 5.8. Schematic illustrating the typical flow path from raw data to chromatography reports, listing the processes and methods used.

handling systems from the strip chart recorder to the client-server data network. The reader is referred to a more comprehensive synopsis of its development and status elsewhere.[14] In today's laboratories, analysts tend to spend as much time and effort in front of a data system terminal as a chromatograph.

While the functional details of individual data systems vary greatly, Figure 5.8 shows the typical key processes or functions (integration, calibration, and quantitation) to convert sample chromatograms (raw data from the detector) into useful chromatography reports. These processes are controlled by a set of user-specified methods residing in the data system as summarized in Table 5.1.

"Integration" is the process of converting digital chromatography raw data into peak data (series of peak retention times and peak areas). In a traditional integration algorithm, the baseline and component peaks are established by monitoring the slope of the raw data and comparing them with a "threshold" to determine the "peak start." A series of integration events in the processing method can be used to customize this integration process. A new class of integration algorithm uses the second derivative of the raw data for peak detection is available in some data systems (e.g., Waters EmPower and Shimadzu Class VP) and appears to be superior for complex chromatograms with sloping baselines (Figure 5.9).[14]

"Calibration" is the process of establishing a calibration curve of the specified analyte from a set of injected calibration standard solutions. Figure 5.10

Table 5.1. Data System Method Types and their Primary Functions

Data system method type	Primary functions
Instrument method	Controls and documents parameters of pump and detector • Pump: flow, gradient conditions, degasser, (oven) • Detector: wavelength, bandwidth, filter response, sampling rate
Sample set or sequence method	Controls and documents parameters of autosampler and sample/standard information • Injection sequence: vial #, inject volume, # injections, run time • Functions: inject samples, equilibrate, calibrate, quantitate • Standard/sample info: name, amount, sample weight, label claim, level
Processing method	Controls and documents integration parameters, component names, calibration and quantitation information • Integration: threshold, peak width, minimum peak area, integration events • Component table: component names, retention times, response factors
Report method	Performs automated custom calculations, formats, and prints reports

shows a typical calibration curve plotting the peak area of the analyte against the amount injected. The response factor (R_f) can be calculated from the slope of the curve or by dividing the peak area with the amount according to the equation below. In "quantitation," peaks in the unknown samples are identified by comparing them with retention times of the component list in the processing method. The amount of the sample can then by calculated by dividing the peak area by its respective response factor.

$$Response\ Factor\ (R_f) = \frac{Area_{Std}}{Amount_{std}} = \frac{Area_{sample}}{Amount_{sample}}$$

"Report" is the process of generating a formatted chromatography report from the result files according to the report method. There are two types of reports—sample and summary reports. Examples are shown in Figures 5.11 and 5.12. The sample report typically documents the entire sample and method information, the sample peak result table, and the chromatogram, and might include additional data such as spectral info from the PDA. The summary report summarizes data of specific sample results and might include sums,

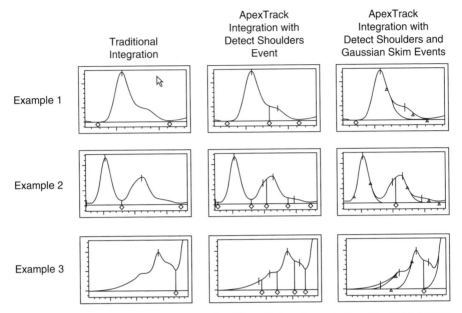

Figure 5.9. Diagrams illustrating the use of traditional integration and "ApexTrack" integration algorithm (Waters EmPower) based on second derivative peak detection. Note that traditional integration can typically handle complex chromatograms well after some manipulations. Diagram courtesy of Waters Corporation.

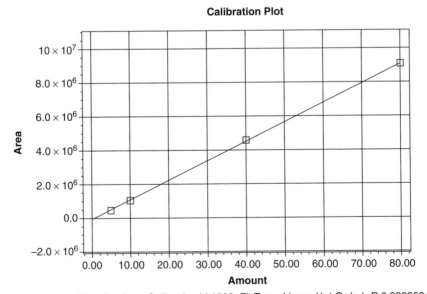

Name: Ethyl Paraben; Calibration Id 1502; Fit Type: Linear (1st Order); R 0.999950; R^2 0.999900

Figure 5.10. Typical calibration curve using external standardization (linear calibration through origin). Note that the slope of the curve is equal to the response factor of the analyte.

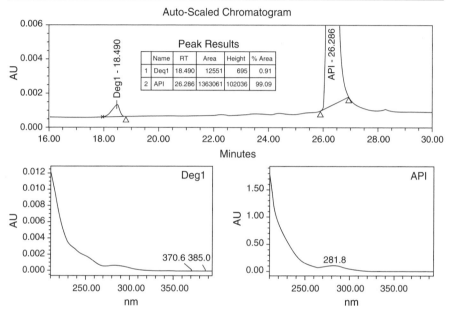

Figure 5.11. A sample report documenting sample info, peak results, chromatogram, and spectral data.

SYSTEM SUITABILITY RESULTS

Name: API

	Sample Name	Name	Vial	Inj	RT	Area	USP Tailing	N	Signal_to_Noise_All	ResultId	USP Resolution
1	sst-nxm	API	2	1	26.458	1393113	1.2	12466	13180.4	2509	23.2
2	sst-nxm	API	2	2	26.466	1395572	1.2	122530	10922.5	2508	23.0
3	sst-nxm	API	2	3	26.488	1394045	1.2	122434	11395.5	2507	23.0
4	sst-nxm	API	2	4	26.471	1392981	1.2	122800	9338.3	2506	22.8
5	sst-nxm	API	2	5	26.491	1392397	1.2	122993	9847.0	2505	22.8
Mean					26.475	1393622	1.2	123085	10936.7		22.9
Std. Dev.					0.014	1240					
% RSD					0.05	0.09					

Figure 5.12. A summary report for system suitability testing documenting result i.d., precision, and other system suitability parameters (S/N ratio, tailing factors, plate count, and resolution). Note that means and precision of each table column can be automatically calculated.

averages, and RSD of specific analytes. For instance, summary reports are used for summarizing system suitability tests (Figure 5.12), content uniformity assays, and dissolution testing. Customized calculations can be used in both report types. These reports can also be exported to the Laboratory Information Management System (LIMS) or other high-level data base archival systems.

5.5.1 Qualitative Analysis Strategies

In qualitative analysis, the goal is to establish the identity of unknown components in the sample.[2,4,5] In HPLC, peaks are identified by its retention time as compared with those of the known analytes in the standard solution. Since absolute retention times are affected by many parameters, relative retention times (RRT) (retention time ratio against a reference component) are often used in more complex samples. Spiking the sample with known analytes is another useful technique to establish peak identity. Other techniques include:

- Peak ratio from two detection wavelengths from a dual-channel UV/Vis detector
- Retention time from a second HPLC column of different selectivity
- λ_{max} or spectral data from PDA or MS data
- Data from specific detectors (e.g., electrochemical, fluorescence, nitrogen-selective detectors)

5.5.2 Quantitation Analysis Strategies

HPLC quantitation methodologies are similar to those used in other chromatographic techniques and fall into three different categories.[2,4,5] Both peak areas and peak heights can be used, though peak area methods are more popular. Readers are referred elsewhere for details and numerical calculation on these quantitation methodologies. Calculation examples are found in Chapter 6.

- Normalized area percent methods, in which area percentage of each peak is reported and the total area percentage equals 100%, are most often used in impurities testing. For testing of pharmaceutical products, the area percentage of impurity is indexed to the parent drug, where the parent peak is set to 100%. This could be a custom field calculation for some data systems.
- External standardization methods are used for most quantitative assays. Solutions containing known concentrations of reference standards of the analytes are required to calibrate (standardize) the HPLC system. Bracketed standards injected before and after the samples set are preferred for improved accuracy.

• Internal standardization methods are common for bioanalytical analysis of drugs in physiological fluids or complex samples requiring extensive sample work-up to compensate for loss occurring in preparation. The internal standard should have similar structures to the analytes and is added before sample work-up. For UV detection, the internal standard must be resolved from any potential sample components. In most bioanalytical LC/MS assays, isotopically labeled analytes (e.g., deuterated analytes) are commonly used.

5.6 SUMMARY OF HPLC OPERATION

The following is a summary of important procedures, a sequence of events, for HPLC operation:

• Filter and degas mobile phase.
• Prime pump, rinse column with strong solvent, and equilibrate column.
• Purge injector and make sure there are no bubbles in the sampling syringe.
• Perform system suitability test (for regulatory testing, see Chapter 9).
• Analyze samples.
• Process and report data.
• Rinse column and shut down pump and lamp.

5.7 GUIDES ON PERFORMING TRACE ANALYSIS

Trace analysis is always difficult and requires rigorous procedures to eliminate potential contaminations and interferences. This is particularly more challenging for gradient HPLC analysis, where any trace contaminants in the weaker mobile phase (MPA) are concentrated on the head of the column during column equilibration and emerge as ghost peaks in the chromatogram. Common examples of HPLC trace analysis using gradient elution are impurity testing of pharmaceutics and environmental testing of trace pollutants. The reader is referred to books[15] and research articles[16] on this topic. This section summarizes guidelines to minimize some of the difficulties and to avoid potential pitfalls:

• Use high-purity reagents. Wear gloves and rinse all glassware.
• Run a procedure blank to ensure that the blank chromatogram does not contain any interfering peaks (see Figure 5.13 for comparison between a good and a bad blank chromatogram in gradient HPLC).

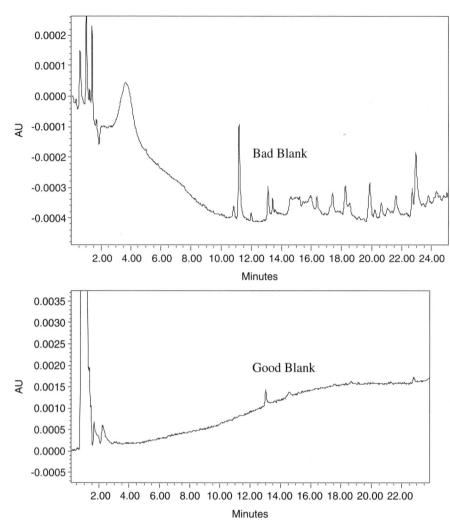

Figure 5.13. A good vs. a bad blank chromatogram from a gradient trace analysis for impurity testing of pharmaceuticals. The "ghost" peaks from the blank injection are derived mostly from the trace contaminants in the weaker mobile phase, which are concentrated during column equilibration. Reprint with permission from reference 16.

- The purity of mobile phase A (MPA, the weaker initial mobile phase) is critical since several milliliters of MPA are used to equilibrate the column and any trace contaminants in MPA are concentrated and elute as ghost peaks in the blank chromatogram. One successful strategy is to use high-purity buffer (>99.995%) and to eliminate the buffer filtration step to minimize potential for contamination.

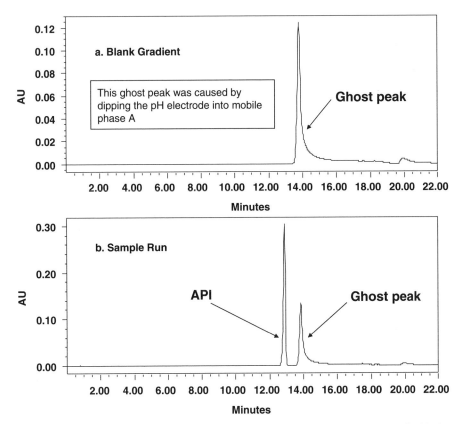

Figure 5.14. (a) Interference peak stemming from contamination of mobile phase A with the pH calibration buffer preservative during pH adjustment procedure by dipping the pH electrode into mobile phase A (the weaker mobile phase). (b) Note that this ghost peak is substantial and elutes close to the active pharmaceutical ingredient (API) in a composite pharmaceutical assay combining assay and impurity testing. This ghost peak is eliminated by not dipping the pH electrode into the mobile phase A during preparation. Reprinted with permission from reference 16.

- The pH adjustment step of MPA might also pose potential problems. The pH calibration buffers often contain a preservative that might contaminate MPA if the pH electrode is dipped during pH adjustment (see Figure 5.14). This problem can be eliminated by pouring out small aliquots into separate containers to check pH.
- HPLC system contamination can also be minimized by:
  — Rinsing column with strong solvents and purging injector.
  — Washing solvent lines with water and organic solvents.
  — Washing system fluidics with 6N HNO_3 after column disconnection if the system is very contaminated.
  — Using a dedicated HPLC system for trace analysis.

- Enhance the sensitivity of the method by:
  - — Selecting a detection wavelength at or near the λ_{max} of the analytes or at far UV (210–230 nm) if the molar absorptivity of the analyte at λ_{max} is too low.
  - — Increasing the injection volume.
  - — Using a detector with higher sensitivity (e.g., a UV/Vis detector typically has higher sensitivity than a PDA detector) and a flow cell with a longer pathlength if possible.
  - — Reducing detector noise by using higher filter response or by increasing the spectral bandwidth of the PDA detector.
- Minimize gradient shifts:
  - — Substantial gradient shift due to absorbance or refractive index changes of the mobile phase during broad gradients. This is particularly serious for running MS-compatible trifluoroacetic acid (TFA) or formic acid gradient at low UV.
  - — Gradient shifts can be reduced by reducing the concentrations of the mobile phase additives (i.e., from 0.5–0.1% TFA to 0.03–0.05%) and by balancing (matching) the absorbances of MPA and MPB. See example in Chapter 8, case study #2. This would also reduce the short-term baseline noise caused by pump blending.

5.8   SUMMARY

In summary, more effective and successful HPLC assays can be accomplished by following some of the "best practices" in mobile phase preparation, strategies in qualitative, quantitative, and trace analysis, and some of the standard operating procedures in HPLC system and data system operation.

5.9   REFERENCES

1. L.R. Snyder and J.J. Kirkland, *Introduction to Modern Liquid Chromatography*, John Wiley & Sons, New York, 1979.
2. L.R. Snyder, J.J. Kirkland, and J.L. Glajch, *Practical HPLC Method Development*, 2nd Edition, Wiley-Interscience, New York, 1997.
3. U.D. Neue, *HPLC Columns: Theory, Technology, and Practice*, Wiley-VCH, New York, 1997.
4. E. Katz, *Quantitative Analysis Using Chromatographic Techniques*, Wiley, Chichester, England, 1987.
5. V.R Meyer, *Practical HPLC*, Wiley, New York. 1988.
6. S. Ahuja and M.W. Dong, eds., *Handbook of Pharmaceutical Analysis by HPLC*, Elsevier, Amsterdam, 2005.

7. HPLC video and slides/tapes training programs, Academy Savant, Fullerton, CA.

8. *Identification and Quantitation Techniques in HPLC, CLC-70* (Computer-based Instruction), Academy Savant, Fullerton, CA.

9. J. Dolan, "HPLC Troubleshooting" columns in *LC.GC* magazine.

10. *Prudent Practices in the Laboratory: Handling and Disposal of Chemicals*, National Research Council, Washington, DC, 1995.

11. A.K. Furr, *CRC Handbook of Laboratory Safety*, 5th Edition, CRC Press, Philadelphia, 2000.

12. M.W. Dong, *Today's Chemist at Work*, **9(8)**, 28 (2000).

13. E. Grushka and I. Zamir, in P. Brown and R. Hartwick, eds., *Chemical Analysis*, **Vol. 98**, Wiley Interscience, New York, 1989, p. 529.

14. R.P. Mazzarese, in S. Ahuja and M.W. Dong, eds., *Handbook of Pharmaceutical Analysis by HPLC*, Elsevier, Amsterdam, 2005.

15. S. Ahuja, *Trace and Ultratrace Analysis by HPLC*, Wiley, New York, 1991.

16. M.W. Dong, G. Miller, and R. Paul, *J. Chromatog.* **987**, 283 (2003).

6

PHARMACEUTICAL ANALYSIS

135

6.1   INTRODUCTION

6.1.1   Scope

This chapter reviews the use of HPLC in pharmaceutical analysis from drug discovery to quality control. The focus is on HPLC analysis of drug substances (DS) and products (DP) such as assay for potency, purity evaluation, and dissolution testing. A case study of the various HPLC methods used during early clinical development illustrates the versatility of this technique. Detailed descriptions of HPLC applications in pharmaceutical development and LC/MS analysis in drug discovery and bioanalytical studies can be found elsewhere.[1-6] The regulatory aspects in pharmaceutical testing are covered in Chapter 9.

6.1.2   Overview: From Drug Discovery to Quality Control

HPLC is the dominant technique for pharmaceutical analysis used in research, development, and quality control[7] (Figure 6.1) with the following functions:

- *Drug discovery:* finding new chemical entities (NCE) for adoption as new drug development candidates.

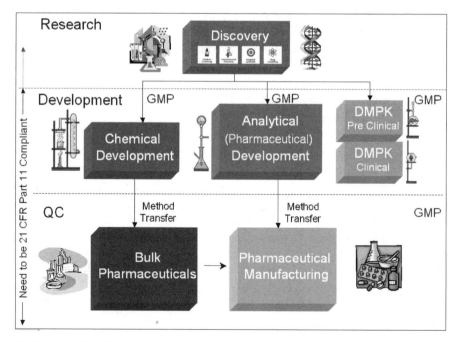

Figure 6.1. Schematic showing the various departments in pharmaceutical research: discovery research, chemical and pharmaceutical development, drug metabolism and pharmacokinetics (DMPK), and manufacturing quality control. Diagram Courtesy of Waters Corporation.

- *Chemical development:* developing viable synthetic routes and scale-up processes for synthesizing active pharmaceutical Ingredients (API).
- *Pharmaceutical development:* developing dosage forms with optimized delivery and stability profiles for clinical supplies and final products.
- *Drug metabolism/pharmacokinetics (DMPK):* evaluating the metabolic and pharmacokinetic profiles of the drug candidates in animal models and human clinical studies.
- *Quality control (QC):* assessing the quality of the final manufactured products against published specification for product release.

Examples of common HPLC applications are listed in Table 6.1. Details on HPLC assays, testing for purity and dissolution performance, bioanalytical analysis, and chiral assays are discussed in later sections. Highlights on sample preparation are reviewed.

6.1.3 Sample Preparation Perspectives in Drug Product Analysis

In general, the "dilute and shoot" approach can be used for most drug substances and parenteral products.[8] A common process of "grind → extract → dilute → filter" is used for most solid dosage forms such as tablets or capsules. More complex dosage forms, such as suppositories, lotions, and creams, and physiological samples (serum or plasma) might require additional sample clean-up and extraction such as liquid-liquid extraction or solid-phase extraction (SPE).[12] The output of pharmaceutical sample preparation is typically an HPLC vial containing the final testing solution with the extracted analytes ready for HPLC analysis. Further details on sample preparation of pharmaceutical products can be found elsewhere.[8]

6.1.4 High-Throughput LC/MS in Drug Discovery Support

In drug discovery, the goal is to discover new chemical entities (NCE) that are safe and efficacious in treating various diseases.[9,10] A modern approach is to use parallel synthesis to create combinatorial libraries of a large numbers of compounds, which are subsequently screened to allow selection of a few drug development candidates with the appropriate pharmacological properties. Since hundreds and thousands of these compounds are created, high-throughput screening (HTS) procedures are required to enhance productivity.[11,12] In the past decade, LC/MS has become one of the most versatile tools for HTS in modern drug discovery processes, particularly in the support of organic synthesis and bioanalytical testing. The organic synthesis support includes confirmation of the synthesis of target compounds, as well as their purity estimation and purification by HPLC. The latter often uses the MS signal as a specific trigger for fraction collection. The bioanalytical support includes quantitation of target analytes in biological matrices in support of preclinical lead optimization in the area of absorption, distribution, metabolism, and excretion

Table 6.1. Examples of Common HPLC and LC/MS Analysis in Pharmaceutical Development

Process	Analytical testing	Analytical technique	HPLC detection
Drug discovery	Purification and characterization of combinatorial libraries and lead compounds, ADME/PK screening	HPLC	MS, MS/MS, PDA, ELSD
Chemical development	Assays and purity evaluation of key raw materials and intermediates during synthesis and process development	HPLC, GC, NMR, IR	PDA, MS
Formulation development and stability evaluation	Identification, assay, purity testing, dissolution, content uniformity, blend uniformity, and cleaning validation	HPLC, IR, UV	PDA, UV, MS
Metabolism and pharmacokinetics	Metabolic profiling and bioanalytical assays of animal models of clinical samples	HPLC, GC	MS/MS, UV, ECD
Quality control	Release testing of final products: assay, identification, purity testing, dissolution, and content uniformity	HPLC, GC, UV	UV, MS

ADME/PK = adsorption, distribution, metabolism and excretion / pharmacokinetic; IR = infrared spectroscopy; PDA = photodiode array detector; NMR = nuclear magnetic resonance; ELSD = evaporative light scattering detector; ECD = electrochemical detector.

(ADME).[11] Figure 6.2 shows a schematic diagram of an LC/MS system configured for high-throughput purification. Figure 6.3 shows examples of fast gradient chromatograms (evaporative light scattering detector, ELSD, and UV 214 nm) of a library compound obtained during purification. Further details in this area are available elsewhere.[12]

6.2   IDENTIFICATION

Identification testing is used to confirm the presence of the API in samples of drug substances or drug products. In most cases, two independent tests are

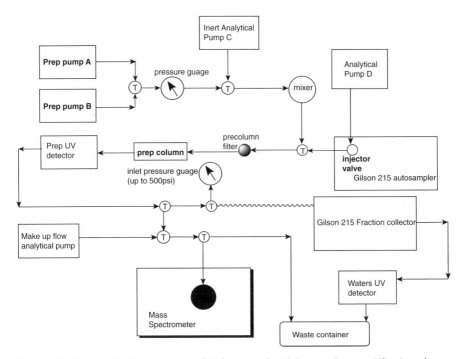

Figure 6.2. Schematic diagram of an LC/MS system for high-throughput puritification of combinatorial library compounds using the MS signal as trigger for fraction collection. Reprinted with permission from reference 12.

required. These include an HPLC retention time match and a spectroscopic test, such as UV or IR spectral match, usually against a qualified reference standard. Identification testing is only performed in critical tasks such as product release or stability evaluation as well as after-market support such as analyzing customer complaints or counterfeit samples.

6.3 ASSAYS

Since most drugs are chromophoric (exhibit UV absorbances), HPLC with UV detection is commonly used in assays for drug potency of drug substances and products. Typical method attributes, requirements, specification limits, and acceptance criteria are listed Tables 6.2 and 6.3. Assay methodologies of many common drug substances and products are published in the United States Pharmacopeia (USP) or European Pharmacopeia (EP) monographs. Assays for potency are performed for product release and stability evaluation.

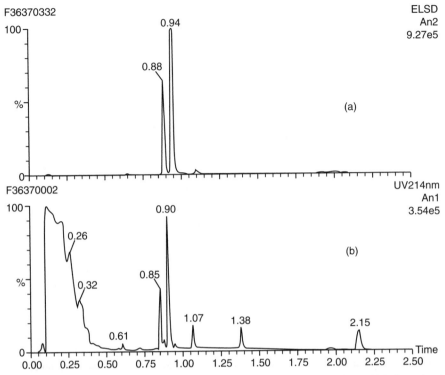

Figure 6.3. Example chromatograms of two target compounds isolated by the high-throughput purification system shown in Figure 6.2. Chromatograms illustrate the performance of fast gradient analysis (ELSD and UV detector trace before purification). Reprinted with permission from reference 12.

6.3.1   Drug Substances

Table 6.2 summarizes typical assay method attributes for drug substances.

6.3.2   Drug Products

The assay of drug product potency is the quantitative determination of the API to ensure conformance to label claim (e.g., the potency for a standard aspirin tablet is 325 mg). Typical specification limits for most drug products is 90–110% of their label claims. For common solid dosage forms such as tablets or capsules, a composite assay of 10–20 units is typically used to minimize tablet-to-tablet variation. Quantitative extraction of the API from the solid formulation is critical. While extraction of the intact dosage form is possible, most products might require some grinding to facilitate quantitative extraction.[8] Alternately, a portion of the composite ground powder equal to the average tablet weight (ATW) may be extracted and assayed to reduce the amount of extraction solvents used. Extraction is mostly conducted in volu-

Table 6.2. Typical Method Attributes for HPLC Methods for Assays of Drug Substances

Method	HPLC with UV detection
Sample preparation	Dissolve drug substances in volumetric flasks at 0.01–1 mg/mL
Quantitation	External standardization using a well-qualified reference standard (e.g., USP reference standard or a working reference standard). Correction for moisture and residual solvents maybe required.
Typical specification limits	98%–102% purity on a dried basis
Method validation acceptance criteria*	Accuracy: three levels in triplicate (70, 100, and 130% of nominal assay concentration); <±2% from true value.
	Precision: six determinations at 100% level; <±2.0 RSD.
	Specificity: API resolved from major impurities, target $R_s > 1.5$.
	Linearity: five levels between 50 and 150% ($r > 0.999$) with % y intercept NMT 2.0%.
	Robustness: Method accuracy unaffected by method perturbations of key parameters.

r = Coefficient of linear correlation, NMT = Not more than. *Typical values cited as examples.

Table 6.3. Typical Method Attributes for HPLC Methods for Assays of Drug Products

Method	Isocratic (or gradient) HPLC with UV detection.
Sample preparation	"Grind → extract → dilute → filter" of 10–20 units
Quantitation	External standardization using a well-qualified reference standard.
Typical specification limits	90–110% of label claim.
Method validation acceptance criteria*	Accuracy: three levels in triplicate (70,100, and 130%); <±2% from spiked value.
	Precision: six determinations at 100% level; <±2.0 RSD.
	Specificity: API resolved from major impurities/degradant. Noninterference from placebo.
	Linearity: five levels between 50 and 150% ($r > 0.999$) with %y intercept NMT 2.0%.
	Robustness: Method accuracy unaffected by slight perturbations.

r = Coefficient of linear correlation, NMT = Not more than. *Typical values cited as examples.

metric flasks and can be facilitated by vortexing, mechanical shaking, or ultrasonication. A two-step extraction process using different extraction solvents might be required for many controlled-release products to extract the API from the polymer matrix.[8] The nature of the extraction solvents (percentage organic solvent, pH, etc.) and extraction time are optimized during method development. The extracts are typically filtered with syringe membrane filters directly into HPLC vials. Table 6.3 summarizes typical HPLC assay method attributes for solid drug products.

An equation showing the calculation of percentage label claim is shown below.

$$\% \text{ LC} = \frac{A_{samp}}{A_{STD}} \times \text{Std. conc.} \times \frac{\text{Vol}_{samp}}{N_{tab}} \times \frac{100}{\text{Label Claim}},$$

where $\text{Std. conc.} = \dfrac{\text{Wt Std(mg)}}{\text{Stock Vol(mL)}} \times \text{Purity} \times \text{CF}$

% LC   = Percentage label claim of active pharmaceutical ingredient (API) per tablet or capsule

A_{samp} = Area of the API peak in the sample solution

A_{std} = Area of the API peak in the standard solution

CF   = Weight Conversion Factor if the API is a different salt form than that of the reference standard

Purity = Purity of the reference standard after correction for moisture, residual solvents, and actual purity of the compound

Vol_{samp} = Volume of the sample solution

N_{tab} = Number of tablets or capsules tested

100   = Conversion to %

6.3.3   Content Uniformity

According to the USP,[16] a content uniformity test is similar to the composite assay test, except assay of an individual tablet/capsule is conducted to ensure processing consistency. Generally, 10 single-tablet assays are required. A test of weight variation can be substituted for a content uniformity assay if the product contains >50 mg of an API and comprises >50% by weight of the dosage form unit. Typical acceptance criteria proposed for uniformity of dosage units are that each of the dosage units must lie between 85.0% and 115.0% of the label claim with an RSD NMT 6.0%. If these criteria are not met, testing of additional tablets is required according to the USP.

Figure 6.4 shows an example of an HPLC assay of a single API product (capsule) with the content uniformity assay results summarized in Table 6.4.

6.3.4   Products with Multiple APIs and Natural Products

The sample preparation and HPLC analysis are more elaborate for formulations with multiple APIs (e.g., over-the-counter (OTC) products) or with natural products. Examples of HPLC analysis of two OTC multi-vitamin products are shown in Figure 6.5, with a summary of method performance for both water-soluble and fat-soluble vitamins[17] listed in Table 6.5. Other examples of HPLC analysis of extracts of natural products (white and red ginseng)[18] are

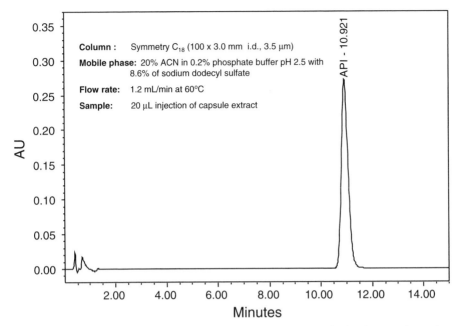

Figure 6.4. Example of an HPLC assay for potency (% label claim) of a drug product using reversed-phase ion pair chromatography. The same method is used for content uniformity testing.

Table 6.4. An Example of Content Uniformity Assay Report

	Sample name	Result ID	% API
1	LIMS#1234#1	1507	99.5
2	LIMS#1234#2	1508	101.0
3	LIMS#1234#3	1509	97.3
4	LIMS#1234#4	1510	102.9
5	LIMS#1234#5	1511	102.4
6	LIMS#1234#6	1512	98.6
7	LIMS#1234#7	1513	100.2
8	LIMS#1234#8	1514	105.5
9	LIMS#1234#9	1515	98.6
10	LIMS#1234#10	1516	101.4
Mean			**100.7**
Min			**97.3**
Max			**105.5**
%RSD			**2.41**

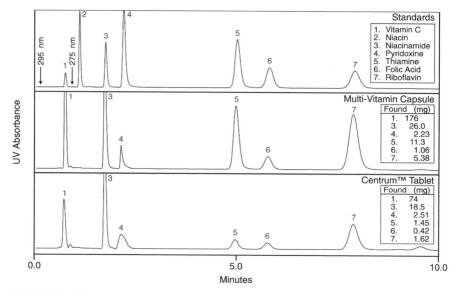

Figure 6.5. HPLC chromatograms showing the analysis of water-soluble vitamins in multi-vitamin capsules tablets. Reprinted with permission from reference 17.

Table 6.5. Summary of Analytical Method Performance Parameters for Vitamin Analysis

Performance parameter	Water-soluble vitamins	Fat-soluble vitamins
Precision		
Retention time	<0.3%	<0.15%
Peak area	<1.0%*	<1.1%
Assay (tablet)	<3% (10% for vitamin C)	<3%
Accuracy (recovery of	90–100%*	98% vitamin A
spikes)		87% vitamin E
Limit of detection	0.6–1.2 ng	<1 ng (vitamins A and E)
(LOD) (S/N = 3)		<10 ng (vitamin E)
Specificity	P.I. < 1.1, Rs > 3	P.I. < 1.1, Rs > 5
	N > 6,000 (confirmation	N > 8,000
	λ_{max} and purity index)	
Range	1–>1,000 ng	1–>10,000 ng (vitamin A)
		5–>20,000 ng (vitamin E)
Linearity	$r > 0.9995$	$r > 0.999$
Ruggedness		
Mobile phase prep	±5% of organic solvent	±3% of the organic solvent
	(R_s of key analytes	(R_s of key analytes
	maintained)	maintained)
Column lifetime	>1,000 injections	>1,000 injections
Column compatibility	Fair	Excellent
Temperature variations	Sensitive	20–45°C

Note: PI = purity index, r = coefficient of linear regression, R_s = resolution, N = column plate count, S/N = signal-to-noise ratio. *Except for vitamin C, which is unstable in water. Reprinted with permission from reference 17.

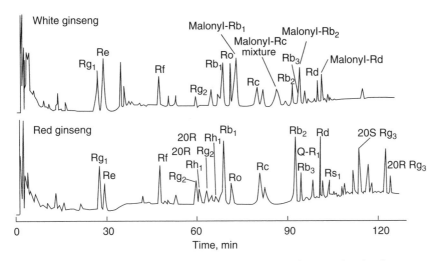

Figure 6.6. HPLC chromatograms of extracts of red and white ginsengs showing the presence of various active ingredients in the complex natural product extracts. Reprinted with permission from reference 18.

shown in Figure 6.6. Note that HPLC assays of natural products typically require high-resolution gradient methods as well as automated summation of all active ingredient peaks to calculate potency.

6.3.5 Assay of Preservatives

A preservative is a substance that extends the shelf-life of drug products by preventing oxidation or inhibiting microbial growth.[14] Preservatives must be monitored in the products since they are considered to be active components. Generic HPLC assays are typically developed for preservatives such as buty-lated hydroxytoluene (BHT), an antioxidant for solid dosage forms, and antimicrobials such as parabens, sodium benzoate, or sorbic acid in liquid for-mulations. For these additive components, typical assay specifications are 85–115% of label claim.

6.4 IMPURITY TESTING

The monitoring of process impurities and degradants is important to assure drug purity and stability. The Food and Drug Administration (FDA) has published updated guidelines, which outline the need for stability testing on drugs and biologics for use in humans.[19] In addition, the International Conference on Harmonization (ICH) has recently revised guidelines for the validation of analytical methods to measure and to report impurities and degradants in new chemical entities.[20,21] For sample preparation, the complete extraction of the parent drug and impurities, and the elimination of contaminants and artifact peaks, are more critical since any unknown peaks above the ICH guideline of ~0.1% might require reporting, identification, or qualification depending on daily dose.[20] A method sensitivity target is often set at limit of quantitation, or LOQ, <0.03% during early analytical development. Impurities with geno-toxicities might have significantly lower specifications (e.g., <10 ppm) and require a much lower method LOQ. Table 6.6 summarizes typical method attributes for impurity testing methods. Methods for drug substances should separate all process impurities. Methods of drug products, however, must separate all degradants from the API and other formulation components (e.g., excipients and preservatives). Impurities are not monitored in drug product methods since they are controlled in the drug substances. Residual solvents are also important impurities in drug substances and are handled mostly by GC.

Figure 6.7 shows an example of an HPLC analysis of impurity testing of a drug product stored for 6 months under accelerated stability conditions (40°C/75% RH). Note that to maximize sensitivity of the trace impurity peaks,

Table 6.6. Typical Method Attributes for HPLC Methods for Impurity Testing of Drug Substances and Products

Method	Gradient (or isocratic) HPLC with UV detection
Sample preparation	Dissolve drug substance at ~0.1–5 mg/mL
	"Grind → extract → dilute → filter" of 5 dosage units for products
Method requirements	Drug substances: must resolve all impurities and APIs
	Drug products: must resolve all degradants and APIs
	Limit of quantitation (LOQ) <0.03–0.1%
Quantitation	Area normalization against the API
	External standardization using well-qualified impurity reference standards
Reporting, identification, and qualification thresholds	Typically 0.05–1.0% according to ICH guidelines and dependent upon maximum daily dose
Typical specification limits	Drug substances: <0.2–1.0% of each specified impurity
	Drug products: <0.2–1.0% of each specified degradant
	(Total impurity/degradant limits are also specified)
Method validation acceptance criteria	Accuracy: three levels in triplicate (LOQ – 1%), <±10–20% of spike values
	Precision: six replicate at LOQ, <10–20% RSD
	Specificity: resolve key specified impurities and degradants
	Linearity: five levels between LOQ – 2% (r < 0.98)

LOQ = limit of quantitation.

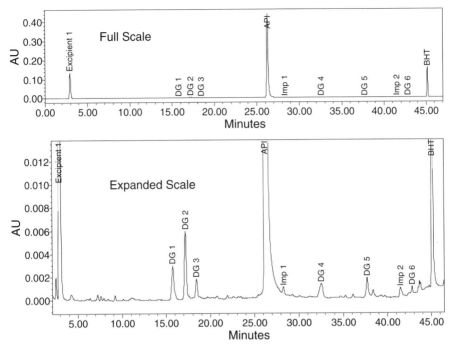

Figure 6.7. An example of impurities testing using gradient HPLC of a pharmaceutical product stored under accelerated stability conditions, noting the presence of excipient, preservative (butylated hydroxytoluene, BHT), impurities (Imp), and degradants (DG). HPLC conditions: column: Waters XTerra MS_{18}, $150 \times 3\,mm$ i.d., $3\,\mu m$; mobile phase: (A) $16\,mM$ ammonium bicarbonate, pH 9.1; (B) acetonitrile; gradient: 3% B to 45% B in $25\,min$; flow rate: $0.8\,mL/min$ at 50°C; detection: $280\,nm$. Additional information can be obtained from reference 22.

relatively large amounts of the sample are typically loaded. However, if area normalization is used, as often is the case in early development, the absorbance of the API peak must not exceed the linearity of the detector (typically 1.0–1.5 absorbance units). It is also important to not to overload the column or the buffering capacity of the mobile phase. Substantial efforts are expended to ensure there are not impurity peak eluting under the API peak as further discussed in Chapters 8 and 9. A typical equation used for calculating the percentage degradant in drug products using the area normalization method is shown below.

$$\text{Degradant (\%)} = \frac{\text{Area of degradant}}{\text{Area of parent API}} \times \frac{100\%}{\text{RRF of degradant}}$$

Where RRF is the relative response factor of the degradant against the API at the monitoring detection wavelength.

6.4.1 Trends in Impurity Testing

Impurity testing of pharmaceuticals is one of the most difficult HPLC method development tasks because of their requirements for both high-resolution and trace analysis in additional to stringent regulatory and reporting guidelines.[20,21] There are several recent trends for these methods:[22,23]

- Gradient methods: to increase peak capacity and enhance detection of highly retained impurities such as dimmers.
- MS-compatible methods: to facilitate identification of unknowns often encountered during process development and early stability studies.
- Composite assay and impurity testing methods[1]: combining both assay and impurity testing in a single method to facilitate testing during early development.
- Stage-appropriate methods[1]: Developing and validating several impurity methods during the drug development cycle. Gradient, high-resolution, and multiple orthogonal methods during early development, which evolve into a single isocratic, robust method to monitor only key degradants for quality control.

6.5 DISSOLUTION TESTING

Dissolution testing is used to measure the release of the API under standardized conditions such as those specified in the United States Pharmacopoeia (USP) using the Apparatus Type I (basket method) or Type II (paddle method) or other apparatus types.[16] This *in vitro* evaluation can be compared or correlated with the *in vivo* bioavailability of the solid dosage formulation from clinical studies. Dissolution testing is performed to check product consistency during formulation development, stability studies, and product releases.[24,26] Since most drugs are chromophoric, the analytical techniques for dissolution testing are UV spectrometry and HPLC. The advantages of HPLC over UV are that HPLC provides higher specificity and sensitivity, indication of stability, and applicability to analyze formulations with multiple APIs or low doses.[26] Typically, the isocratic HPLC assay methods for potency determination are adopted and validated for dissolution testing. Table 6.7 lists typical HPLC dissolution method attributes. Figure 6.8 shows examples of the dissolution profiles of a controlled release product generated during a formulation development study. These results were used to optimize the formulation.

6.6 CLEANING VALIDATION

Cleaning validation tests are performed to ascertain the effectiveness of the procedure used to clean pharmaceutical processing equipment (e.g., blender,

Table 6.7. Typical Method Attributes for HPLC Methods for Dissolution Testing

Method	Isocratic HPLC with UV detection
Sample preparation	None
Quantitation	External standardization
Typical specification limits	>Q + 5% in 30–60 min for immediate release products (Stage 1) ±10–15% of release specification at specified time points for controlled release products
Method validation acceptance criteria	Accuracy: three levels in triplicate at specification levels; <±5% of true values
	Precision: 12 sample units NMT 10–20% RSD
	Specificity: API resolved from major impurities/degradants and product components
	Linearity: Linearity established between 10 and 150% of analyte concentration, $r > 0.99$ with % y-intercept NMT 5.0%

Q is a value specified in USP <711> in USP28-NF23. NMT = Not more than.

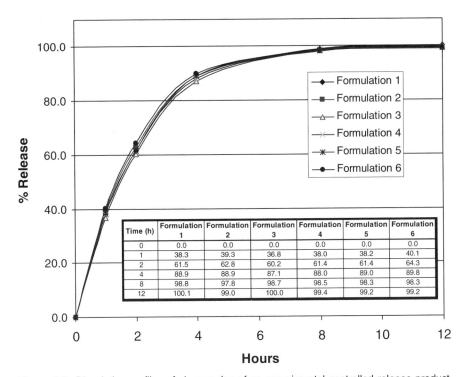

Figure 6.8. Dissolution profiles of six samples of an experimental controlled release product, manufactured under different processing conditions. The inset shows the actual data obtained.

Table 6.8. Typical Method Attributes for HPLC Methods for Cleaning Validation

Method	Isocratic (or gradient) HPLC with UV detection
Sample preparation	Extraction of wipers or swabs
Quantitation	External standardization
Typical specification limits	<0.1% of maximum daily dosage for product residues
Method validation acceptance criteria	Accuracy: three levels in triplicate at LOQ to 20 X LOQ or the acceptance criterion; <±50% of the spiked values
	Precision: six replicates at LOQ; NMT 20% RSD
	Specificity: API resolved from other components in a procedural blank sample
	Linearity: five levels from LOQ to 20 X LOQ; $r > 0.98$

tablet press).[27] Cleaning validation methods are developed to check for analytes such as product residuals, degradants, cleaning reagents, or other contaminants. Sampling is typically performed by swiping a specific location with a wiper or a swab, which is then extracted for analysis by a number of analytical techniques including HPLC, total organic carbon, or spectrophotometric techniques such as UV or bioassay for microbial contaminations. Sensitivity and freedom from interfering peaks are the most critical requirements for cleaning validation methods (Table 6.8).

6.7 BIOANALYTICAL TESTING

Bioanalytical testing is the analysis of drugs and their metabolites in physiological fluids (plasma, serum, tissue extracts), typically from experimental animals or human subjects during preclinical or clinical studies.[28] In the last decade, conventional HPLC methods using UV, fluorescence, or electrochemical detection after elaborate sample clean-up have now been mostly replaced by LC/MS/MS methodologies. Sample preparation may include off-line procedures such as protein precipitation, liquid-liquid extraction, or solid-phase extraction (SPE).[29] Alternately, samples can be cleanedup online using turbulent flow chromatography (using a small column packed with large particles at high flow rate to eliminate the serum proteins). The 96-well microplate is becoming the standard format for conducting these assays in the high-throughput mode. To minimize assay variability (between samples or between batches) associated with ionization efficiency of MS and extraction recovery during sample preparation, the use of stable isotope-labeled internal standards (H^2, C^{13}, or N^{15}) in quantitative bioanalysis is gaining widespread acceptance, especially in the late stage of drug development. For higher sensitivity and selectivity, a daughter ion from the fragmentation of the parent analyte is used for quantitation—a technique called selective reaction monitor (SRM). Table 6.9

Table 6.9. Typical Method Attributes for LC/MS/MS Methods for Bioanalytical Testing

Method	Isocratic or gradient HPLC with MS/MS detection
Sample preparation	SPE, protein precipitation, liquid-liquid extraction, or online clean up; typically using the 96-well microplate format
Quantitation	Internal standardization based on a stable isotope-labeled analog of the analyte is prefered
Method validation acceptance criteria	Accuracy: three levels, five determinations each; <15–20% of the spiked values
	Precision: three levels, five determinations each; NMT 15–20% RSD, intraday and interday precision
	Specificity: analytes resolved from other metabolites or endogenous components; check blank samples from six subjects for interference.
	Linearity: Regression coefficient of calibration curve >0.95 based on 6–8 levels covering the entire dynamic range
	Stability: room temperature, frozen storage, freeze/thaw, in biological matrices (high and low level in triplicates). Stability of standard solution and stability of samples in autosampler also required).

SPE = solid-phase extraction.

summarizes the typical method attributes of LC/MS/MS methods for bioanalytical testing.

6.8 CHIRAL ANALYSIS

Chirality is a subject of intense interest in the pharmaceutical industry since many drug molecules have asymmetric centers and can exist as enantiomers with markedly different biological activities. Today, most new drugs and those under development consist of a single optically active isomer. Thus, during the synthesis of enantiomerically pure drugs, chiral separation of drug molecules and of their precursors has become an active research area in the past decade. Chiral separation is typically performed on an HPLC column with a chiral-specific phase (CSP) or capillary electrophoresis with a chiral-specific eluent. Common CSPs include low-molecular-weight selectors (Pirkle type), macrocyclic selectors (cyclodextrins, crown-ethers, macrocyclic antibiotics), and macromolecular chiral selectors (proteins, molecular imprinted polymers, polysaccharides).[30,31] Figure 6.9 shows a chiral HPLC separation using a macrocyclic antibiotic column under reversed-phase conditions. This type of chiral column appears to work well for a large number of pharmaceuticals.[31] Figure 6.10 shows a chromatogram of a Zolmitriptan tablet extract spiked with 0.15% of the (R)-enantiomer and another impurity at the 1% level. A detailed description of chiral separation and method development guidelines can be found elsewhere.[32]

Column: CHIROBIOTIC T column (250 x 4.6mm)

Mobile ph.: 70% methanol, 30% water,

Flow rate: 0.5 mL/min @ ambient temp.

Detection: UV at 210 nm

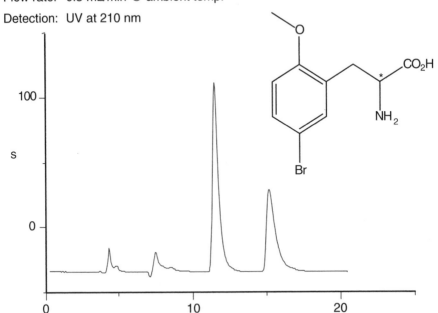

Figure 6.9. An example of a chiral separation using a macrocyclic chirobiotic column of un-natural amino acids as chiral synthesis intermediates. Courtesy of Chirotech Technologies, UK.

6.9   CASE STUDY: HPLC METHODS IN EARLY DEVELOPMENT

A case study involving the analysis of various samples during the early development of a drug candidate is included here to illustrate the application of the myriad HPLC methods in these processes. The drug substance is a neutral molecule with very low aqueous solubility. Two key degradation products (DP1 and DP2) are its isomerization and hydrolysis products. Figure 6.11 shows the chromatogram of the drug substance using an initial HPLC method developed during drug discovery. This method is a composite (combined) method for both assay and impurity testing.

Though several deficiencies were found for this initial method, it was judged adequate for early formulation development. A gradient method with improved resolution of all impurities was developed later (see Chapter 8, Case Study 1). A decision was made to formulate a suspension dosage form for early clinical studies.

Figure 6.12 shows the dissolution HPLC chromatogram of the product in 3% sodium dodecylsulfate (SDS) at pH 6.8. SDS was required in the media

Column: Daicel Chiralpak AD-H (250 x 4.6 mm i.d.)
Mobile ph.: Hexane, I-PrOH, MeOH, Diethylamine
(75:10:15:0.1)
Flow rate: 1.0 mL/min @ 25 °C
Detection: 225 nm

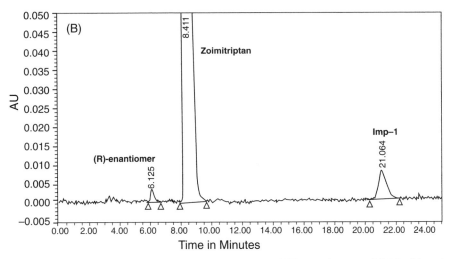

Figure 6.10. Zolmitriptan formulation spiked with 0.15% of (R)-enantiomer and 1.0% of Imp-1 in the developed method. Reprinted with permission from reference 32.

due to the low solubility of the API. Note the elution of the paraben preservative in the dissolution chromatogram, which is separated from the other components. Figure 6.13 shows the LC/MS/MS chromatograms of the daughter ion (M-18) of the tetra-deuterated internal standard (upper trace) and the drug parent (lower trace) in a plasma sample used for generating bioavailability profiles of each clinical study subject. Daughter ions are typically used for LC/MS/MS quantitation because better signal-to-noise performance can be achieved due to higher specificity.

6.10 SUMMARY

This chapter provides an overview of modern HPLC application in pharmaceutical development from drug discovery to quality control of the final product. The HPLC applications discussed include assays for potency and content uniformity, testing for purity, dissolution performance, and cleaning validation. Typical method attributes for both drug substance and drug products are described. In addition, other HPLC and LC/MS applications in drug discovery, bioanalytical assays, and chiral separation are reviewed. A case study illustrating the various HPLC methods used during early development of a new drug substance is described.

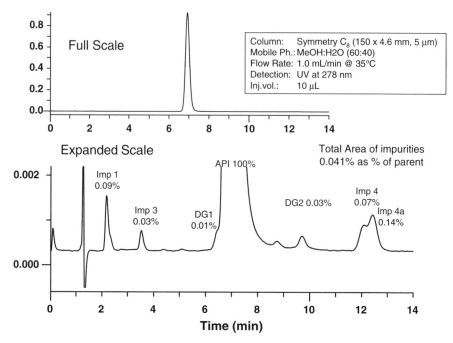

Figure 6.11. An example of a composite assay (combining both assay and impurity testing in one method) for a drug substance in early development. Note that the absorbance of the API must be <1.5 absorbance units (AU) to prevent detector saturation. Since the method has some deficiencies (e.g., partial resolution between several peaks), an improved gradient method was thus developed (Chapter 8).

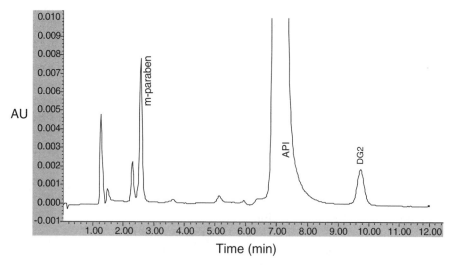

Figure 6.12. Dissolution chromatogram of the drug product (suspension) using the same assay method as in Figure 6.8, showing the presence of the API, preservatives, and degradant.

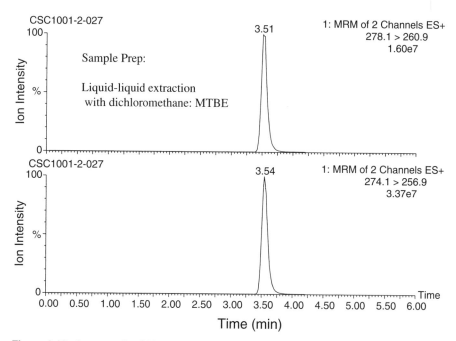

Figure 6.13. An example of bioanalytical analysis using LC/MS/MS analysis for the quantitation of bioavailability clinical samples. Sample preparation used liquid-liquid extraction of dichloromethane and t-butylmethylether. A tetra-deuterated analog of the API was used as an internal standard (upper trace) for the quantitation of the actual API of the subject sample (lower trace). Note that a fragmented daughter ion. (M/z = M − 18, 257) was used for the actual quantitation of the API with excellent sensitivity (low ng/mL) and selectivity.

6.11 REFERENCES

1. S. Ahuja and M.W. Dong, eds., *Handbook of Pharmaceutical Analysis by HPLC*, Elsevier, Amsterdam, 2005.

2. J.M. Miller and J.B. Crowther, *Analytical Chemistry in a GMP Environment: A Practical Guide*, Jossey-Bass, New York, 2000.

3. S. Ahuja and S. Scypinski, eds., *Handbook of Pharmaceutical Analysis*, Academic Press, New York, 2001.

4. S. Ahuja, *Impurities Evaluation of Pharmaceuticals*, Marcel Dekker, New York, 1998.

5. G. Lunn and N.R. Schmuff, *HPLC Methods for Pharmaceutical Analysis*, Wiley-Interscience, New York, 1997.

6. W.R. LaCourse, *Anal. Chem.*, **74**, 2813 (2002).

7. R. Mazzarese, in S. Ahuja and M.W. Dong, eds., *Handbook of Pharmaceutical Analysis by HPLC*, Elsevier, Amsterdam, 2005, Chapter 20.

8. C. Choi and M.W. Dong, in S. Ahuja and M.W. Dong, eds., *Handbook of Pharmaceutical Analysis by HPLC*, Elsevier, Amsterdam, 2005, Chapter 5.

9. W.P. Janzen, *High Throughput Screening: Methods and Protocols* (Methods in Molecular Biology, 190), 1st Edition, Humana Press, New York, 2002.

10. R. Kong, in S. Ahuja and M.W. Dong, eds., *Handbook of Pharmaceutical Analysis by HPLC*, Elsevier, Amsterdam, 2005, Chapter 17.

11. J.N. Kyranos, et al., *Current Opinions in Biotechnology*, **12**, 105 (2001).

12. W. Goetzinger, et al., *International J. of Mass Spectrometry*, **238**, 153 (2004).

13. J.T. Carstensen and C.T. Rhodes, *Drug Stability: Principles and Practices, 3rd Edition*, Marcel Dekker, New York, 2000.

14. A. Wong and A. Delta, in S. Ahuja and M.W. Dong, eds., *Handbook of Pharmaceutical Analysis by HPLC*, Elsevier, Amsterdam, 2005, Chapter 13.

15. ICH Harmonized Tripartite Guideline—Stability Testing of New Drug Substances and Products Q1A (R2), 2003.

16. United States Pharmacopeia and National Formulary, USP 27-NF 22, United States Pharmacopoeial Convention, Inc., Rockville, MD, 2005.

17. M.W. Dong and J.L. Pace. *LC.GC*, **14(9)**, 794 (1996).

18. O. Stieher, CHEMTECH, **28(4)**, 26–32 (1998).

19. FDA Guidelines for Center for Drug Evaluation and Research (CDER), http://www.fda.gov/cder

20. ICH Guidelines for Impurity Testing: Q3B(R) Impurities in New Products, 2003, http://www.ich.org/pdfICH/Q3BRStep4.pdf

21. ICH Guidelines for Impurity Testing: Q3A(R) Impurities in New Drug Substances, 2003, http://www.ich.org/pdfICH/Q3ARStep4.pdf

22. M.W. Dong, G. Miller, and R. Paul, *J. Chromatog.* **987**, 283–290 (2003).

23. H. Rasmussen et al., in S. Ahuja and M.W. Dong, eds., *Handbook of Pharmaceutical Analysis by HPLC*, Elsevier, Amsterdam, 2005, Chapter 6.

24. J.B. Dressman, *Pharmaceutical Dissolution Testing*, Marcel Dekker, New York, 2005.

25. Q. Wong and V. Gray, in S. Ahuja and M.W. Dong, eds., *Handbook of Pharmaceutical Analysis by HPLC*, Elsevier, Amsterdam, 2005, Chapter 15.

26. M.W. Dong and D.C. Hockman, *Pharm. Tech.*, **11**, 70 (1987).

27. A. Plasz, in S. Ahuja and M.W. Dong, eds., *Handbook of Pharmaceutical Analysis by HPLC*, Elsevier, Amsterdam, 2005, Chapter 16.

28. I.D. Wilson, *Bioanalytical Separations (Handbook of Analytical Separations, Volume. 4)*, Elsevier Science, Amsterdam, 2003.

29. D.A. Wells, *High Throughput Bioanalytical Sample Preparation: Methods and Automation Strategies*, Elsevier Science, Amsterdam, The Netherlands, 2003.

30. Y.V. Heyden, et al., in S. Ahuja and M.W. Dong, eds., *Handbook of Pharmaceutical Analysis by HPLC*, Elsevier, Amsterdam, 2005, Chapter 18.

31. T.E. Beesley and R.P.W. Scott, *Chiral Chromatography*, John Wiley & Sons, New York, 1999.

32. M.K. Srinivasu, et al., *J. Pharm. Biomed. Anal.*, **37(3)**, 453 (2005).

7

FOOD, ENVIRONMENTAL, CHEMICAL, AND LIFE SCIENCES APPLICATIONS

Modern HPLC for Practicing Scientists, by Michael W. Dong
Copyright © 2006 John Wiley & Sons, Inc.

7.1   INTRODUCTION

7.1.1   Scope

This chapter describes the use of HPLC in food, environmental, chemical (polymer, ion chromatography), and life sciences applications. The HPLC advantages, method requirements, and trends of these application areas are summarized with descriptions of the methodologies of key analytes. References for further studies are listed.

7.2   FOOD APPLICATIONS

HPLC is widely used in food analysis in product research, quality control, nutritional labeling, and residual testing of contaminants. HPLC is a cost-effective technique ideally suited to the testing of labile components in complex matrices.[1,2] Tables 7.1 and 7.2 summarize the key analytes, advantages, and trends in food analysis by HPLC. Analysis of natural components (sugars, fats, proteins, amino acids, and organic acids), food additives (preservatives, colors, flavors, and sweeteners), and contaminants are discussed. Readers are referred to books.[1-4] review and research articles,[5] and other sources such as methodologies from the Association of Official Analytical Chemists (AOAC

Table 7.1. Summary of Food Applications by HPLC

Key applications	Natural components, additives, and contaminants
HPLC advantages	High-resolution, multi-component quantitative analysis
	Sensitivity and specific analysis using diverse detectors
	Amenable to labile components
Trends	Increasing use of ELSD for nonchromophoric analytes and
	LC/MS/MS for residual trace analysis

ELSD = Evaporative light scattering detector.

Table 7.2. Summary of HPLC Modes and Detection in Food Analysis

Food analysis	HPLC mode	Detection
Natural food components		
Carbohydrates	Mixed mode, RPC, IEC	RI, PAD, ELSD
Lipids, triglycerides, and cholesterols	NARP, NPC	RI, UV, ELSD
Fatty acids and organic acids	IEC, RPC	UV, RI
Proteins, peptides	IEC, RPC, SEC	UV
Amino acids	IEC, RPC	UV/Vis*, FL*
Food additives		
Acidulants, sweeteners, flavors,	RPC	UV
Anitoxidants and preservatives	RPC	UV
Colors and dyes	RPC	UV/Vis
Vitamins	RPC	UV, FL
Contaminants		
Mycotoxins (aflatoxins)	RPC	FL, UV
Pesticide and drug residues	RPC	UV, FL, MS/MS
PAHs and nitrosamines	RPC	UV, FL, MS/MS

RPC = reversed-phase chromatography, IEC = ion-exchange chromatography, NPC = normal-phase chromatography, SEC = size-exclusion chromatography, NARP = nonaqueous reversed-phase, PAD = pulsed amperometric detector, ELSD = evaporative light scattering detector.
*Pre-column or post-column derivatization required.

International) or the American Society of Testing and Materials (ASTM) for further details on specific applications. Recent regulations in food labeling have increased the level of food testing.[6]

7.2.1 Natural Food Components

7.2.1.1 Sugars Common carbohydrates are monosaccharides (glucose, fructose), disaccharides (sucrose, maltose, lactose), trisaccharides (raffinose), and polysaccharides (starch). HPLC offers a direct, quantitative method for simple sugars, which requires a specialty cationic resin-based column and refractive index or evaporative light scattering detection.[7,8] UV detection at low wavelengths (195 nm) can be used but is more prone to interferences.[8]

Different resin columns (Ca^{++}, Pb^{++}, H^+, Ag^+, etc.) are used for different sample types, with calcium being the most common form and water as the mobile phase at 85°C[8] (see Figure 7.1). The basis of separation is mixed-mode, which includes size-exclusion, ion-exclusion, ligand exchange, and hydrophobic interaction with the polystyrene support. This reliable assay methodology has excellent precision, sensitivity (10–20 ng), linearity (0.050–800 μg), and column lifetime.[8] This is the preferred approach for most routine assays. An amino column with acetonitrile/water (80/20) (AOAC method) can also be used, however, sensitivity is lower and column lifetime is limited.[7]

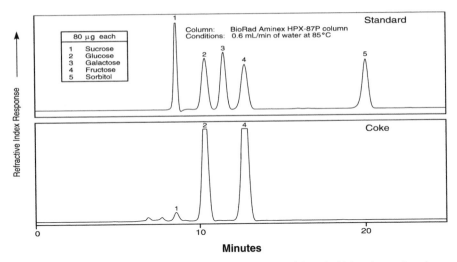

Figure 7.1. HPLC analysis of carbohydrates (simple sugars) in soft drink using resin column and refractive index detection. HPLC conditions: column: BioRad Aminex HPX-87C (300 × 7.8 mm i.d.); injection volume: 10 μL; mobile phase: water, flow rate: 0.6 mL/min at 85°C; detection: refractive index. Chromatogram courtesy of PerkinElmer, Inc.

A highly sensitive approach developed in the 1990s utilizes high-performance anion-exchange (HPAE) using a high-pH eluent with pulsed-amperometric detection.[7] At high pH, carbohydrates become charged and are separated as the oxyanion. Pulsed amperometric detection (PAD) offers low-picomoles sensitivity and is compatible with gradient elution. This approach is particularly valuable in biotechnology applications such as the characterization of carbohydrate moieties in glycoproteins.

7.2.1.2 Fats, Oils, and Triglycerides

Fats and oils are triesters of glycerol with fatty acids (triglycerides). Fatty acids are commonly analyzed as their methyl esters by GC after sample transesterification. Free acids can also be analyzed by HPLC with RI or low UV detection.[9] Phospholipids (lecithin, cephalin, or phosphotidyl-inositol) from soybeans are used as food additives. Common sterols are cholesterols and phytosterols (sitosterols and stigmasterol). These sterols can readily be separated on silica columns with detection at 210 nm[10] (Figure 7.2).

HPLC analysis of triglycerides is simple and quantitative. It can be used for quality control and product testing for adulteration. Edible oils can be injected directly as a 5% solution in acetone using nonaqueous reversed-phase (NARP) chromatography with two columns packed with 3-μm C18 particles and a mobile phase of acetone/acetonitrile with refractive index detection[11] (Figure 7.3). Gradient elution yields better separation of mono- and diglycerides and can also be used with evaporative light scattering detection (ELSD) or UV detection at 210–220 nm. Note that UV detection is less

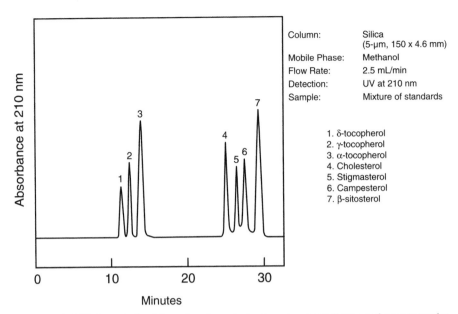

Figure 7.2. HPLC analysis of tocopherols and sterols using normal-phase chromatography. Reprinted with permission from reference 9.

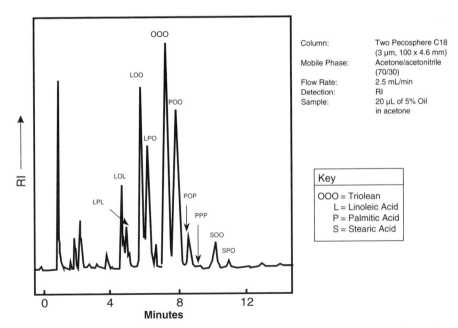

Figure 7.3. HPLC analysis of triglycerides in olive oil using nonaqueous reversed-phase chromatography with refractive index detection. Reprinted with permission from reference 11.

reliable as it is more subject to interferences. Another viable approach for high-resolution triglyceride analysis is high-temperature capillary GC (using metal-clad fused silica columns) with a programmable-temperature vaporization (PTV) injector.

7.2.1.3 *Free Fatty Acids and Organic Acids*   Free fatty acids can be readily analyzed by RPC with an acidified mobile phase and detection at 210 nm.[9] However, pre-column derivatization of free fatty acids to form derivatives such as p-bromophenacyl esters, 2-nitrophenyl hydrazides, or anthrylmethylesters, can improve the chromatographic performance and detection sensitivity by UV or fluorescence detection.[9] Nevertheless, GC of their methyl esters remains the more common assay method.

Organic acids occur naturally in foodstuffs as a result of biochemical metabolic processes, hydrolysis, or bacterial growth.[12] Common organic acids are:

- Unsubstituted—formic, acetic, propionic, butyric
- Substituted—glycolic, lactic, pyruvic, glyoxylic
- Di- or Tricarboxylic—oxalic, succinic, fumaric, maleic, malic, citric acid

They are added as stabilizers and preservatives (sorbic acid) or for endowing flavor, taste, or aroma (propionic acid, citric acid). Some organic acids are indicators for ripeness, bacterial activity, or spoilage. Organic acids in wines and juices are best analyzed by HPLC with hydrogen resin-based columns with RI or UV detection at 210 nm[13] (Figure 7.4). Since these columns also separate sugars and alcohols, they are particularly useful in the monitoring fermentation products during wine making.

7.2.1.4 *Proteins and Amino Acids*   Total protein in food and feed samples is commonly determined by Kjeldahl (acid digestion/titration) or Dumas (pyrolysis) or elemental analysis.[14] HPLC can separate major proteins and furnish protein profiles and speciation information. HPLC can be used to further characterize specific proteins via peptide mapping and amino acid sequence analysis. HPLC modes used for protein include IEC, SEC, RPC, and affinity chromatography with typical UV detection at 215 nm or MS analysis. Details on protein separations are discussed in the life sciences section.

Amino acid analysis is useful for assessment of the nutritional value of food and feed.[15] The essential amino acids are methionine, cysteine, lysine, threonine, valine, isoleucine, leucine, phenylalanine, tyrosine, and tryptophan. Glutamic acid is a commercial flavor enhancer. Since amino acids are nonchromophoric (possess low UV absorbance), either pre-column or post-column derivatization is typically needed:

- Post-column derivatization—IEC is used to separate free amino acids followed by ninhydrin or o-phthaldehyde (OPA) post-column derivatiza-

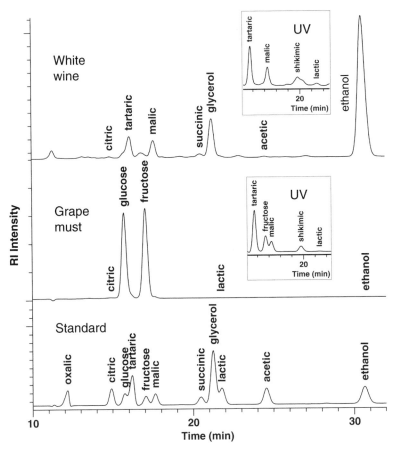

Figure 7.4. HPLC analysis of organic acids in white wine and grape must using a resin column with refractive index and UV detection at 210nm. HPLC conditions: two Polypore H (10 μm, 220 × 4.6mm); mobile phase: 0.01N H_2SO_4; flow rate: 0.2mL/min at 60°C. Reprinted with permission from reference 13.

tion to form chromophoric derivatives. This technique is used in dedicated amino acid analyzers, preferred for most food testing. An example of the ion-exchange chromatogram with post-column derivatization is shown in Figure 1.7.

• Pre-column derivatization—RPC analysis of phenylisothiocyanate (PITC), o-phthaldehyde (OPA), 9-fluorenylmethyl chloroformate (FMOC), or other derivatives of amino acids with UV or fluorescence detection is most common. This is the preferred methodology for life science research because of its higher sensitivity. More examples of pre-column derivatization of amino acids are shown in the life science section of this chapter.

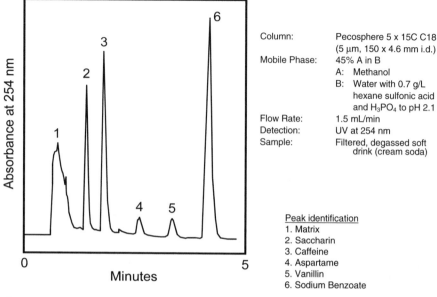

Column: Pecosphere 5 x 15C C18 (5 µm, 150 x 4.6 mm i.d.)
Mobile Phase: 45% A in B
A: Methanol
B: Water with 0.7 g/L hexane sulfonic acid and H_3PO_4 to pH 2.1
Flow Rate: 1.5 mL/min
Detection: UV at 254 nm
Sample: Filtered, degassed soft drink (cream soda)

Peak identification
1. Matrix
2. Saccharin
3. Caffeine
4. Aspartame
5. Vanillin
6. Sodium Benzoate

Figure 7.5. HPLC analysis of sweetener, flavors, and preservative in a soft drink sample using gradient reversed-phase chromatography and UV detection. Chromatogram courtesy of PerkinElmer, Inc.

7.2.2   Food Additives

The analysis of food additives is important for nutritional labeling and quality control.[16] Common additives are:

- Acidulants
- Sweeteners—aspartame, saccharin (Figure 7.5)
- Colors and food dyes (Figure 7.6)
- Vitamins—water-soluble and fat-soluble vitamins (Figure 7.7)
- Flavor compounds—bitter, pungent, aromatic compounds, and flavor enhancers (Figure 7.8)
- Antioxidants and preservatives

Selected examples in the quality control of beverages and baby formula are shown in Figures 7.5–7.7. Figure 7.5 shows an ion-pair gradient HPLC analysis of a cream soda sample showing the presence of sweetener (saccharin, aspartame), preservative (sodium benzoate), flavor (vanillin), and additive (caffeine). Figure 7.6 shows the HPLC analysis of various food colors using UV detection at 290 nm (a universal wavelength for these colors). Figure 7.7 shows the HPLC analysis of various water-soluble vitamins in a fortified drink and a baby formula sample. The HPLC analysis of multi-vitamins[17] is also covered in Chapter 6 in the assay of multi-vitamin tablets. A case study on the

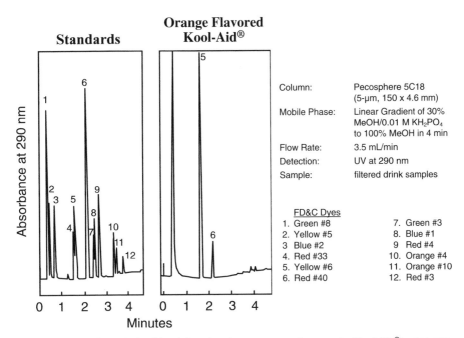

Figure 7.6. HPLC analysis of food dyes in a beverage powder sample (Kool-Aid®) using gradient reversed-phase chromatography and UV detection at 290 nm, which can detect many different colored dyes. Chromatogram courtesy of PerkinElmer, Inc.

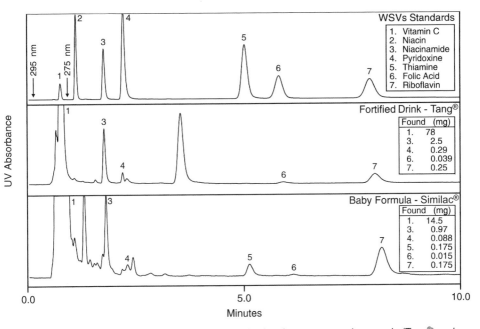

Figure 7.7. HPLC analysis of water-soluble vitamins in a beverage powder sample (Tang®) and a baby formula (Similac®) using ion-pair reversed-phase chromatography and UV detection. Reprinted with permission from reference 17.

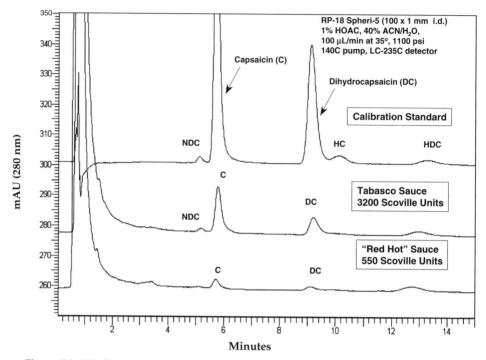

Figure 7.8. HPLC analysis of capsaicins, the principle components responsible for the "heat" in hot pepper, in two samples of hot sauces using reversed-phase chromatography with acidified mobile phase and UV detection. Reprinted with permission from reference 18.

analysis of capsaicins (Figure 7.8) illustrates the versatility of HPLC analysis and customized reporting in food testing.

7.2.2.1 Flavors: A Case Study on HPLC Analysis of Capsaicins

GC is used mostly in the analysis of fragrances and flavors due to their volatility. One notable exception is the analysis of capsaicins, the components responsible for the "heat" in hot pepper.[18] The degree of pungency of hot peppers differs significantly due to the varying concentrations of capsaicins in different species (Table 7.3). In addition, the "hotness" of each capsaicin also varies widely as expressed in millions of Scoville units: nordihydrocapsaicin (NDC, 9.3), capsaicin (C, 16.1), dihydrocapsaicin (DC, 16.1), homocapsaicin (HC, 6.9), and homodihydrocapsaicin (HDC, 8.1).

The traditional way to evaluate the "hotness" in peppers or hot sauces is through a taste panel, measured by the "Scoville" scale, through a series of comparative evaluations of diluted samples. The taste panel method is subjective, costly, and time consuming. Capsaicins can be analyzed directly (without derivatization) by HPLC. The method is precise and rapid (20 minutes), and requires only a drop of filtered hot sauce. Figure 7.8 shows the

Table 7.3. The Pungency Levels of Various Peppers

Pepper	Pungency (in Scoville units)
Bell pepper	0
Anaheim	250–1,400
Jalapeno	3,000
Hungarian yellow	4,000
Japanese chili	20,000–30,000
Tabasco	30,000–50,000
Cayenne	50,000–100,000
Indian birdeye	100,000–125,000
Japanese kumataka	125,000–150,000
Habanero	300,000

Reprinted with permission from reference 18.

Table 7.4. Customized Report on the "Hotness" of Tabasco Sample

Capsaicin	Amount		Scoville units	
	ng	% of capsaicin in sauce	Of pure component $\times 10^6$	Of component in sample
Nordihydrocapsaicin (NDC)	3.37	0.001	9.3	78
Capsaicin (C)	50.1	0.013	16.1	2020
Dihydrocapsaicin (DHC)	24.4	0.006	16.1	984
Homodihydrocapsaicin (HDC)	10.0	0.003	8.1	204
Total	88.0	0.022		3286

Reprinted with permission from reference 18.

HPLC separation of individual capsaicins in the calibration standard and two hot sauces. Best of all, the degree of "heat" of each sample can be automatically reported in Scoville units by summing up the percentage of each capsaicin multiplied by its Scoville score (Table 7.4).

7.2.3 Contaminants

The analysis of food contaminants, in particular any toxic or biologically active residue, is important for public health or quality control reasons.[19] Examples are mycotoxins (aflatoxins) and pesticide and drug residues. Sample preparation is typically elaborate and might involve deproteinization, solvent extraction, and clean-up via solid-phase extraction (SPE). The use of highly sensitive and specific LC/MS/MS is increasing and has simplified some of the sample preparation procedures.

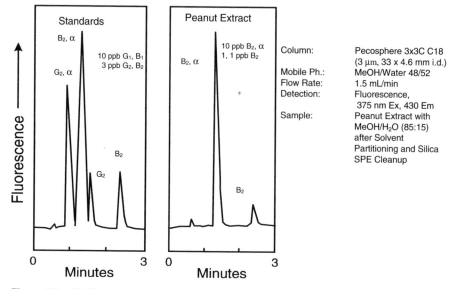

Figure 7.9. HPLC analysis of aflatoxins in a peanut extract sample using reversed-phase chromatography and fluorescence detection. Chromatogram courtesy of PerkinElmer, Inc.

7.2.3.1 *Mycotoxins*
Details on the analysis of mycotoxins are found elsewhere.[19] Examples of mycotoxins are:

- Aflatoxins—toxic metabolites from molds (Figure 7.9)
  —B1, B2, G1, G2 (in grains and nuts) or M1, M2 (in dairy products)
- Ochratoxins A, B, and C—fungal metabolites found in maize, wheat, or oats
- Zearalenone—found in maize
- Citrinine—from molds found in grains
- Trichothecenes—fungal poisons found in grains
- Patulin—found in apple juice

7.2.3.2 *Antimicrobial Additives*
Common antimicrobials and their HPLC detection wavelength[20] are listed below. Details can be found elsewhere.[20]

- Tetracycline antibiotics   (UV 355 nm)
- Beta-lactam antibiotics   (UV 210 nm)
- Polyether antibiotics   (FL)
- Aminoglycoside antibiotics   (FL)
- Macrolide antibiotics   (UV 231 nm)
- Nitrofurans   (UV 362 nm)

* Sulfonamides   (UV 275 nm)
* Quinoxaline 1,4-dioxide   (UV 350 nm)

7.2.3.3 Pesticide Residues Pesticide residue analysis often requires elaborate sample preparation procedures and analyte enrichment, which might involve extraction (homogenization, solvent extraction), clean-up (solvent partitioning, GPC, SPE), and analyte derivatization.[21] Examples are shown in the environmental applications section. Common pesticide residues found in food crops, vegetables, or fruit are:

* Insecticides—carbamates, organophosphates
* Fungicides—benzimidazoles, thiophanates, dithiocarbamates
* Herbicides and growth regulators—ureas, phenylureas, triazines, pyridazines, glyphosate, diquat/paraquat

7.3 ENVIRONMENTAL APPLICATIONS

Environmental testing is performed by environmental testing labs, water authorities, or utilities, government and state health labs, and agrochemical or pesticide manufacturers.[22–24] HPLC is widely used as exemplified by the adoption of many HPLC methods by the U.S. Environmental Protection Agency (EPA) and other regulatory agencies.[22] Table 7.5 summarizes the common sample types, key analytes, and the HPLC advantages. A brief description and some examples of ion chromatography can be found in a later section.

Table 7.5. Summary of Environmental Applications of HPLC

Samples	Drinking water, wastewater, solid waste, and air samples
Key applications	Pesticides, herbicides and plant growth regulators (carbamates, glyphosate, diquat/paraquat, pyrethrins, organophosphorus compounds)
	Industrial pollutants (dyes, surfactants, amines, PAHs, PCBs, dioxins, phenols, aldehydes)
HPLC advantages	Amenable to thermally labile, nonvolatile, and polar components
	Rapid and automated analysis of complex samples
	Precise and quantitative (with spectral confirmation possible)
	Quantitative recovery (good sample preparation technique)
	Sensitive and selective detection (UV, FL, MS)
	Can tolerate injections of large volumes of aqueous samples (0.5–1 mL)

Table 7.6. Selected EPA Methods Using HPLC

US EPA method	Analyte	Mode	Detection
531.1/8318	Carbamates	RPC	FL, post-column derivatization
547	Glyphosate	IEC	FL, post-column derivatization
549	Diquat/paraquat	RPC	UV
550.1/610/8310/TO13	PAHs	RPC	UV/FL
TO-11	Formaldehyde	RPC	UV, pre-column derivatization

FL = fluorescence

7.3.1   Listing of Important U.S. EPA HPLC Methods

There are more than 40 U.S. EPA methods utilizing HPLC. These are 500 Series methods for drinking water, 600 Series methods for wastewater, 8000 Series methods for solid wastes, and TO methods for toxic organics in ambient air samples.[22] An updated list can be found in the U.S. EPA website (www.epa.gov). Selected examples are listed in Table 7.6.

7.3.2   Pesticides Analysis

Pesticide analysis methods developed by manufacturers, in support of registration petitions to the regulatory agencies, are usually company proprietary. These are mostly HPLC methods since most pesticides are thermally labile and might not survive the GC process. These methods are used to generate assay, stability, residue, and metabolism data. In the United States, regulations require pesticide manufacturers to notify the EPA using a premanufacture notice (PMN), describing structure, impurities, byproducts, environmental fate, and toxicology data. Pesticide determinations are also performed on drinking water, wastewater, and solid waste using EPA methods. Residue analysis of foodstuffs is performed using ASTM, AOAC, or National Pesticides Survey (NPS) methods. Figure 7.10 shows an example of screening of trace levels of pesticides (100–2000 ppt) in drinking water using solid-phase extraction for pre-concentration, RPC gradient separation, and UV detection with photodiode array spectral confirmation.[24] LC/MS/MS is rapidly becoming the method of choice for the determination of trace environmental contaminants.

7.3.2.1   Carbamates and Glyphosate   Carbamates are an important class of insecticides for food crop applications.[25] They are nonvolatile and cannot be analyzed by GC. Carbamate testing (U.S. EPA methods 531.1 and 8318) is performed using post-column derivatization with o-phthaldehyde (OPA) for fluorescence detection. Figure 7.11 shows an example of carbamate analysis used in the United States for monitoring the quality of drinking water.[25] One of the key HPLC advantages for the analysis of drinking water or wastewater

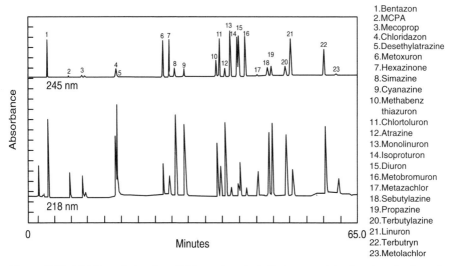

1. Bentazon
2. MCPA
3. Mecoprop
4. Chloridazon
5. Desethylatrazine
6. Metoxuron
7. Hexazinone
8. Simazine
9. Cyanazine
10. Methabenz thiazuron
11. Chlortoluron
12. Atrazine
13. Monolinuron
14. Isoproturon
15. Diuron
16. Metobromuron
17. Metazachlor
18. Sebutylazine
19. Propazine
20. Terbutylazine
21. Linuron
22. Terbutryn
23. Metolachlor

Figure 7.10. HPLC analysis of pesticides using photodiode array detection. Levels of pesticides are 100–2,000 parts-per-trillion preconcentrated by solid-phase extraction. Chromatogram courtesy of PerkinElmer, Inc.

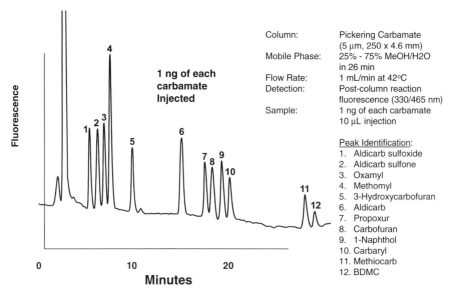

Column: Pickering Carbamate
                   (5 μm, 250 x 4.6 mm)
Mobile Phase: 25% - 75% MeOH/H2O
                   in 26 min
Flow Rate: 1 mL/min at 42°C
Detection: Post-column reaction
                   fluorescence (330/465 nm)
Sample: 1 ng of each carbamate
                   10 μL injection

Peak Identification:
1. Aldicarb sulfoxide
2. Aldicarb sulfone
3. Oxamyl
4. Methomyl
5. 3-Hydroxycarbofuran
6. Aldicarb
7. Propoxur
8. Carbofuran
9. 1-Naphthol
10. Carbaryl
11. Methiocarb
12. BDMC

Figure 7.11. HPLC analysis of carbamates according to U.S. EPA Method 531.1 using post-column reaction and fluorescence detection. Reprinted with permission from reference 25.

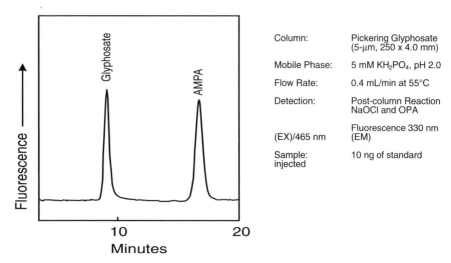

Column:	Pickering Glyphosate (5-μm, 250 x 4.0 mm)
Mobile Phase:	5 mM KH_2PO_4, pH 2.0
Flow Rate:	0.4 mL/min at 55°C
Detection:	Post-column Reaction NaOCl and OPA
(EX)/465 nm	Fluorescence 330 nm (EM)
Sample: injected	10 ng of standard

Figure 7.12. HPLC analysis of glyphosate and aminomethylphosphonic acid (AMPA, the primary degradation product of glyphosate) according to U.S. EPA Method 547 using post-column reaction and fluorescence detection. Chromatogram courtesy of Pickering Laboratories.

is the ability to inject a large volume of aqueous sample (e.g., 1 mL) under reversed-phase conditions to enhance sensitivity.

Glyphosate is a common herbicide that is often monitored in drinking water. Analytical testing is performed by U.S. EPA method 547 using a post-column reaction HPLC system similar to that in carbamate testing (Figure 7.12).

7.3.3   Polynuclear Aromatic Hydrocarbons (PAH)

Polynuclear aromatic hydrocarbons (PAH) are formed by pyrosynthesis during the combustion of organic matter and have widespread occurrence in the environment. PAHs are found as trace pollutants in soil, air particulate matter, water, tobacco tar, coal tar, used engine oil, and foodstuffs such as barbequed meat. Many PAHs are carcinogenic in experimental animals (e.g., benzo(a)pyrene) and are implicated as causative agents in human cancers. Analytical techniques include HPLC and GC. Advantages of HPLC are the ability to resolve isomeric PAHs, and the selective and sensitive quantitation by UV and fluorescence detection. U.S. EPA methods for PAH are 550.1, 610, and 8310.

7.3.3.1 Case Study: Quick Turnaround Analysis of PAHs by HPLC in Multimedia Samples   This case study describes quick turnaround HPLC methodologies for the analysis of 16 priority pollutant PAHs in multimedia samples such as contaminated soil, air particulate matter, used engine oil, and

Table 7.7. Comparison of U.S. EPA Method 8310 and the Quick Turnaround HPLC Method for PAHs in Contaminated Soil

Step	EPA Method 8310	Quick turnaround
Extraction	Soxhlet, EPA Method 3540 or Sonication EPA Method 3530	Sonication (bath type)
Concentration solvent exchange	Kuderna-Danish	None (or nitrogen evaporation)
Clean-up	Silica adsorption EPA Method 3630 or GPC EPA Method 3640	None
HPLC column	PE HCODS Sil-X (10-μm, 250 × 2.6 mm)	PE ChromSpher-3, (3-μm, 100 × 4.6 mm)
Analysis time	45 min	16 min
Detection	UV (254 nm)	UV (280/335/360 nm)
	Fluorescence with emission filter 280 nm (ex)/>390 nm (em)	Fluorescence with six programmed wavelengths
Data handling	UV for the first four peaks	UV for acenaphthylene only
	Fluorescence for 12 other PAHs	Fluorescence for all 15 PAHs
Total analysis time	1–2 days	1 hour

Reprinted with permission from reference 26.

water samples.[26] The use of rapid extraction, direct injection of sample extracts, and improved HPLC separation coupled with more specific programmed fluorescence and UV detection reduces the total assay time from 1 to 2 days for the traditional method to about one hour for the quick turnaround method (see Table 7.7). The method also decreases assay cost and solvent use, and is less demanding on the skill level of the analyst. Method sensitivity was 20 ppb for soil/sediment, low μg/1000 M^3 for air particulate matter, and low ppt levels for water samples. Examples chromatograms of the selective detection of PAHs in a marine sediment and an air particulate matter extract are shown in Figures 7.13 and 7.14. Details of PAH analysis in multimedia samples including sampler preparation procedures are found in reference 26.

7.4 CHEMICAL, GPC, AND PLASTICS APPLICATIONS

HPLC is widely used in the chemical and plastics industries. Applications in the chemical industry are quite similar to those for testing pharmaceutical ingredients. They include assay and purity testing of synthetic chemicals such as raw materials, precursors, monomers, surfactants, detergents, and dyes.[27,28] In the plastics industry, GPC is used for polymer characterization in product research and quality control. RPC is used in the determination of polymer additives.

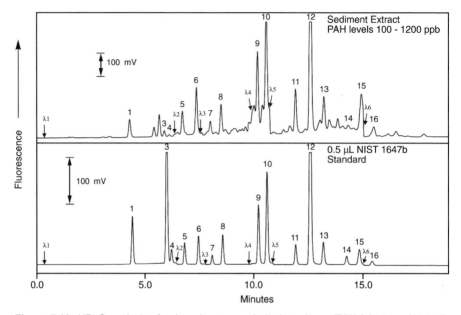

Figure 7.13. HPLC analysis of polynuclear aromatic hydrocarbons (PAHs) in a marine sediment (NIST SRM 1941) using programmed fluorescence detection. Reprinted with permission from reference 26.

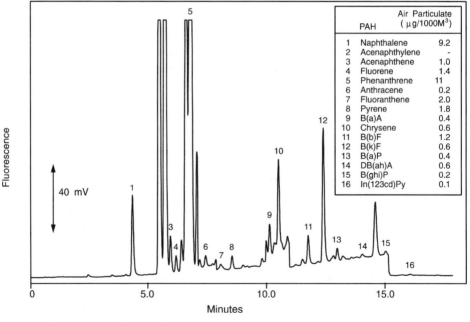

Figure 7.14. HPLC analysis of polynuclear aromatic hydrocarbons (PAHs) in an air particulate extract sample collected from the Long Island Expressway, New York, using programmed fluorescence detection. Reprinted with permission from reference 26.

Table 7.8. Molecular Weight Range of PLgel Columns

Pore size (Å)		Practical linear mol. wt. range
Designation	Actual	
50	10	100–750
100	25	100–2,000
500	40–50	500–30,000
10^3	100	1,000–80,000
10^4	300	5,000–500,000
10^5	1000	50,000–5×10^6
10^6	4000	5×10^5–4×10^7
Mixed-bed B	Mixed	500–8×10^6 (10 μm)
Mixed-bed C	Mixed	500–2×10^6 (5 μm)
Mixed-bed D	Mixed	100–4×10^5 (5 μm)

Columns from Polymer Laboratories. Reprinted with permission from reference 28.

7.4.1 Gel-Permeation Chromatography (GPC) and Analysis of Plastics Additives

GPC is a form of size-exclusion chromatography (SEC) used by polymer chemists and plastics engineers for the characterization of synthetic or natural polymers. Separation is by effective molecular size or hydrodynamic volume using columns packed with materials of 5–10 μm particle size (e.g., cross-linked polystyrene gels) with well-defined pore distribution (see Table 7.8).

A typical GPC system uses an organic solvent as the mobile phase (e.g., THF or toluene) delivered by a precise pump and a refractive index detector.[28] In GPC, large macromolecules are excluded from most intraparticular pores and emerge first. Small molecules penetrate more pores and elute last. Medium-size molecules permeate a fraction of the pores and elute with intermediate retention times according to their molecular weights. GPC is best used for high-molecular-weight components (molecular weight >1,000). All solutes should elute before the sample solvent or any total permeation marker. Any adsorptive interaction of the solute with the packing must be suppressed. GPC has short and predictable retention times but with moderate resolution and limited peak capacity. It is difficult to separate analytes with size differences of less than 10%.

GPC is often used in sample profiling studies shown in an example in Figure 7.15 for use in production quality control of plastics materials. Here, "good and bad" batches of supposedly identical alkyd resins were profiled.[28] The high-molecular-weight components in the "bad" batch were causing processing problems. Molecular weight calibration is not required for this type of application.

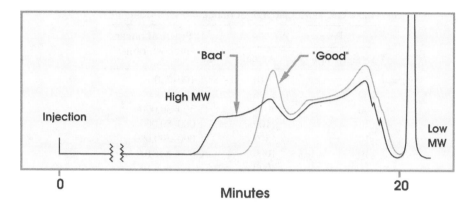

Column:	Two PE PLgel 100Å 10-µm Mixed Bed Columns
Mobile Phase:	THF
Flow Rate:	1 mL/min
Detection:	UV at 254 nm
Sample:	100 µL of a 0.5% solution of alkyl resins

Figure 7.15. Gel-permeation chromatography (GPC) analysis of "good and bad" alkyl resins using UV detection. Reprint with permission from reference 28.

For characterization of molecular weight averages and distribution (MWD), the column set must first be calibrated using a set of molecular weight standards (such as polystyrene or polyethylene glycol, see Figure 1.8). These molecular weight averages and MWD are important because they control the thermal, physical, mechanical, and processing properties of the polymer (e.g., number-average molecular weight (M_n)—tensile strength, hardness; weight-average molecular weight (M_w)—brittleness, flow properties; z-average molecular weight (M_z)—flexibility, stiffness; viscosity-average molecular weight (M_v)—extrudability, molding properties).[27,28] Readers are referred elsewhere for details for these calculations.[27] Specialized GPC software coupled to the chromatography data system is required for these calculations.[28] Figure 7.16 shows a set of molecular weight calibration curves using GPC columns of different pore sizes. Note that mixed-bed columns are particular useful for screening samples of unknown molecular weights or of wide MWD. Figure 7.17 shows a sample of polystyrene foam sample with annotations of various molecular weight averages and polydispersity. If an appropriate set of calibration standard is not available, other techniques such as universal calibration, multiple light scattering detector or mass spectrometry can be used.[27]

Figure 7.18 shows an example of the gradient analysis of polyethylene additives using RPC and UV detection. This application is important in the quality control of additive levels in various plastics.

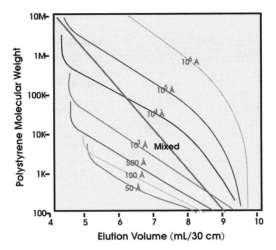

Figure 7.16. Calibration curves of PLgel GPC columns from Polymer Laboratories. See Table 7.7 for column details. Reprinted with permission from reference 28.

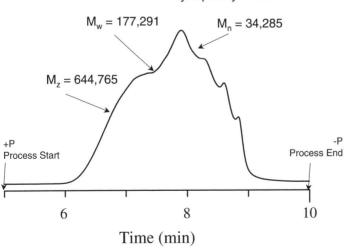

Column:	PLgel Mixed-bed 5-μm
Mobile Phase:	THF
Flow Rate:	0.8 mL/min
Detection:	UV at 270 nm
Sample:	Polystyrene foam sample in THF (10 mg/mL, 50 μL injection)

Polydispersity = 5.17

$M_w = 177,291$ $M_n = 34,285$

$M_z = 644,765$

+P
Process Start

-P
Process End

6   8   10

Time (min)

Figure 7.17. Gel-permeation chromatography (GPC) analysis of a polystyrene foam sample using UV detection. Annotations are calculated molecular weight averages and polydispersity.

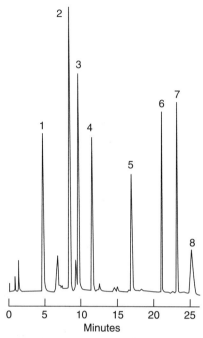

Column: Pecosphere HCODS Cl8
(150 x 2.6 mm, 5 ∝m)

Detection: UV at 200 nm

Mobile Ph.: 50% - 100% CH3CN/H20
Linear gradient in 20 min;

Flow Rate: 1.5 mL/min @ 60°C

Peak identification
1. Tinuvin P
2. Ionol
3. 2-6, di-t-butyl-4-ethyl phenol
4. Oleomide
5. Kemamide-E
6. Irganox 1010
7. Irganox 1076
8. Tris-nonylphenyl phosphate

Figure 7.18. HPLC analysis of polyethylene additive using gradient reversed-phase chromatography and UV detection. Chromatogram courtesy of PerkinElmer, Inc.

7.5 ION CHROMATOGRAPHY

Ion chromatography refers to the analytical technique used in the analysis of low levels of anions and cations in the environmental, chemical, petrochemical, power generation, and electronic industries.[29,30] Ion chromatography is a subset of high-performance ion-exchange chromatography using conductivity detection, often with a suppressor for enhancing sensitivity. Although HPLC equipment is largely used, ion chromatography has evolved as a specialized market segment dominated in the past two decades by a single manufacturer (Dionex). Product innovations from Dionex include enhanced detection using various suppressors, low-capacity latex-based pellicular column packing, and metal-free (PEEK-based) and reagent-free ion chromatography systems. Other equipment manufacturers include Metrohm and Alltech. Figures 7.19–7.22 show examples illustrating the performance of ion chromatography analysis of anions and cations showing the excellent sensitivity and selectivity of this technique.[30] Ion chromatography (IC) is used routinely for environmental testing (e.g., EPA Method 300 for the testing of anions in water) and for monitoring of many manufacturing processes.[29,30]

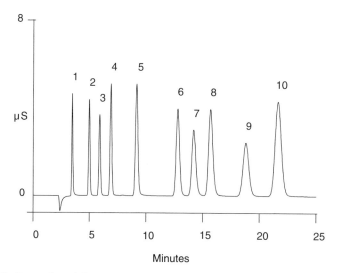

Figure 7.19. Separation of the common anions and disinfectant byproduct anions on IonPac® AS9-HC and AG9-HC columns. Columns: 4 × 250 mm IonPac AS9-HC and 4 × 50 mm AG9-HC. Eluent: 9 mM sodium carbonate. Flow rate: 1 mL·min⁻¹. Injection volume: 25 μL. Detection: Suppressed conductivity utilizing the anion self-regenerating suppressor (4 mm), recycle mode. Ions: 1) fluoride (3 mg·L⁻¹); 2) chlorite (10 mg·L⁻¹); 3) bromate (20 mg·L⁻¹); 4) chloride (6 mg·L⁻¹); 5) nitrite (15 mg·L⁻¹); 6) bromide (25 mg·L⁻¹); 7) chlorate (25 mg·L⁻¹); 8) nitrate (25 mg·L⁻¹); 9) phosphate (40 mg·L⁻¹); 10) sulfate (30 mg·L⁻¹). Reprinted with permission from reference 30.

7.6 LIFE SCIENCES APPLICATIONS

HPLC life science applications focus on the separation, quantitation, and purification of biomolecules such as proteins, peptides, amino acids, nucleic acids, nucleotides, and polymerase chain reaction (PCR) amplification products.[31–34] These are diversified and active research areas in medical research and drug discovery.

7.6.1 Proteins, Peptides, and Amino Acids

Chromatography of proteins is performed by IEC, RPC, and SEC, as well as affinity and hydrophobic interaction chromatography (HIC). Chromatography of large proteins is problematic especially using RPC with silica-based materials for the following reasons[33–35]:

- Hydrophobic and ionic characteristics of proteins, leading to their adsorption in many LC packings with active sites (i.e., acidic silanols)
- Possible conformational changes and denaturing of proteins during the chromatographic process

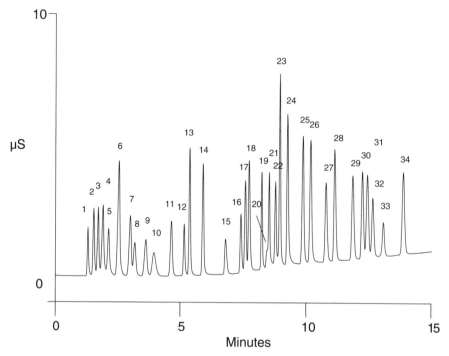

Figure 7.20. Fast gradient separation of inorganic anions and organic acids on an IonPac® AS11 column. Column: 4 × 250 mm IonPac AS11. Eluent: 0.5 mM NaOH for 2 minutes, followed by a linear gradient from 0.5 mM to 5 mM NaOH in 3 minutes and a second linear gradient from 5 mM to 38.25 mM NaOH for the last 10 minutes. Flow rate: 2.0 mL·min⁻¹. Injection volume: 10 μL. Detection: Suppressed conductivity utilizing the anion self-segenerating suppressor (4 mm), recycle mode. Ions: 1) isopropylmethylphosphonate (5 mg·L⁻¹); 2) quinate (5 mg·L⁻¹); 3) fluoride (1 mg·L⁻¹); 4) acetate (5 mg·L⁻¹); 5) propionate (5 mg·L⁻¹); 6) formate (5 mg·L⁻¹); 7) methylsulfonate (5 mg·L⁻¹); 8) pyruvate (5 mg·L⁻¹); 9) chlorite (5 mg·L⁻¹); 10) valerate (5 mg·L⁻¹); 11) monochloroacetate (5 mg·L⁻¹); 12) bromate (5 mg·L⁻¹); 13) chloride (2 mg·L⁻¹); 14) nitrite (5 mg·L⁻¹); 15) trifluoroacetate (5 mg·L⁻¹); 16) bromide (3 mg·L⁻¹); 17) nitrate (3 mg·L⁻¹); 18) chlorate (3 mg·L⁻¹); 19) selenite (5 mg·L⁻¹); 20) carbonate (5 mg·L⁻¹); 21) malonate (5 mg·L⁻¹); 22) maleate (5 mg·L⁻¹); 23) sulfate (5 mg·L⁻¹); 24) oxalate (5 mg·L⁻¹); 25) ketomalonate (10 mg·L⁻¹); 26) tungstate (10 mg·L⁻¹); 27) phthalate (10 mg·L⁻¹); 28) phosphate (10 mg·L⁻¹); 29) chromate (10 mg·L⁻¹); 30) Citrate (10 mg·L⁻¹); 31) tricarballylate (10 mg·L⁻¹); 32) isocitrate (10 mg·L⁻¹); 33) cis-aconitate and 34) trans-aconitate (10 mg·L⁻¹ combined). Reprinted with permission from reference 30.

- Slow diffusivities and sorption kinetics (large van Deemter C term)
- "Restricted" diffusion in small-pore supports leads to band-broadening

As mentioned in Chapter 3, protein separations typically require specialized columns packed with wide-pore polymer supports or silica materials with extra-low silanol activity. Figure 7.23 shows an example of an RPC gradient separation of a protein mixture using a column packed with Vydac C4 bonded

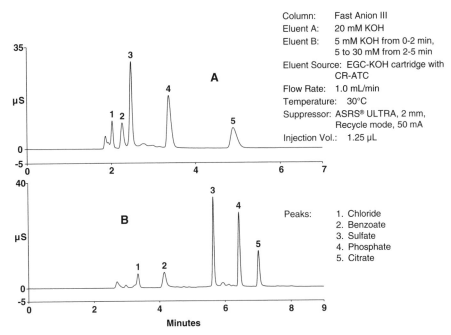

Column: Fast Anion III
Eluent A: 20 mM KOH
Eluent B: 5 mM KOH from 0-2 min,
               5 to 30 mM from 2-5 min
Eluent Source: EGC-KOH cartridge with
               CR-ATC
Flow Rate: 1.0 mL/min
Temperature: 30°C
Suppressor: ASRS® ULTRA, 2 mm,
               Recycle mode, 50 mA
Injection Vol.: 1.25 µL

Peaks: 1. Chloride
          2. Benzoate
          3. Sulfate
          4. Phosphate
          5. Citrate

Figure 7.21. Isocratic (A) and gradient (B) separations of a diet cola sample. Diagram courtesy of Dionex Corporation.

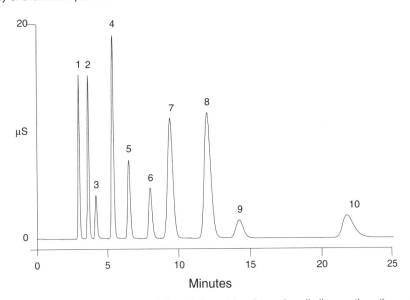

Figure 7.22. Isocratic separation of the alkali metal cations, the alkaline earth cations, and ammonium ion on an IonPac® CS12A column. Column: 4×250 mm IonPac CS12A. Eluent: 18 mM methanesulfonic acid. Flow rate: 1 mL·min^{-1}. Injection volume: 25 µL. Detection: suppressed conductivity utilizing the cation self-regenerating suppressor (4 mm), recycle mode. 1) lithium (1 mg·L^{-1}); 2) sodium (4 mg·L^{-1}); 3) ammonium (5 mg·L^{-1}); 4) potassium (10 mg·L^{-1}); 5) rubidium (10 mg·L^{-1}); 6) cesium (10 mg·L^{-1}); 7) magnesium (5 mg·L^{-1}); 8) calcium (10 mg·L^{-1}); 9) strontium (10 mg·L^{-1}); 10) barium (10 mg·L^{-1}). Reprinted with permission from reference 30.

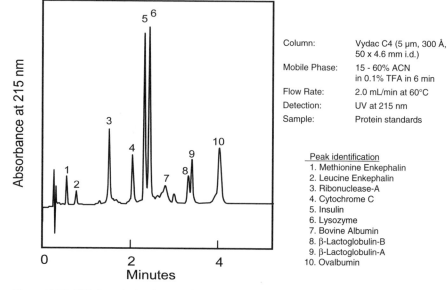

Figure 7.23. HPLC analysis of a protein mixture using reversed-phase chromatography using a silica-based column and UV detection at 215 nm. Reprinted with permission from reference 35.

phase on a wide-pore (300Å) silica support. Vydac columns were the first generation of silica-based columns found in the 1980s to be amenable to RPC of proteins. Gradient elution using 0.1% trifluoroacetic acid (TFA) and acetonitrile is commonly used. Detection at 210–230 nm or 280 nm is typical. The C4-bonded phase is preferred over the more hydrophobic C18 for protein separations.

Polymer support materials such as cross-linked polystyrene-divinylbenzene, polyethers, and polymethacrylates have been used successfully for many years in bioseparations.[31,35] Their strength and performance has improved in recent years though they still lag behind silica in efficiency. Major advantages are the wider useful pH range (1–14) and the absence of active silanol groups. Figure 7.24 shows a RPLC separation of proteins using a polymer column packed with polymeric materials illustrating the excellent peak shape performance of these columns.

Note that small-pore packings are problematic for large biomolecules, which can become entangled or trapped in the pores leading to slower mass transfer and additional band-broadening. Figure 7.25 shows comparative chromatograms of a protein mixture on columns packed with 100 Å and 300 Å polymeric materials, respectively.[35] Large proteins (peaks 5 and 6) show substantial peak broadening in the column packed with the smaller 100 Å support due to "entangled" diffusion into the smaller pores. Thus, the separation of biomolecules is best performed on columns packed with wide-pore

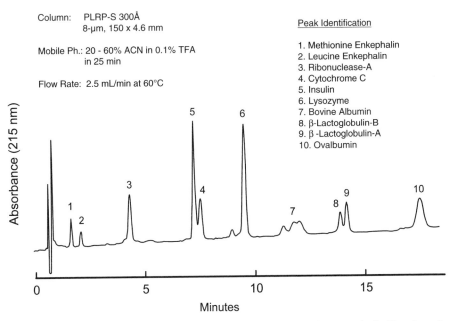

Column: PLRP-S 300Å
              8-µm, 150 x 4.6 mm

Mobile Ph.: 20 - 60% ACN in 0.1% TFA
                in 25 min

Flow Rate: 2.5 mL/min at 60°C

Peak Identification

1. Methionine Enkephalin
2. Leucine Enkephalin
3. Ribonuclease-A
4. Cytochrome C
5. Insulin
6. Lysozyme
7. Bovine Albumin
8. β-Lactoglobulin-B
9. β -Lactoglobulin-A
10. Ovalbumin

Figure 7.24. RPLC chromatogram of a protein mixture using a column packed with polymeric material. Reprinted with permission from reference 35.

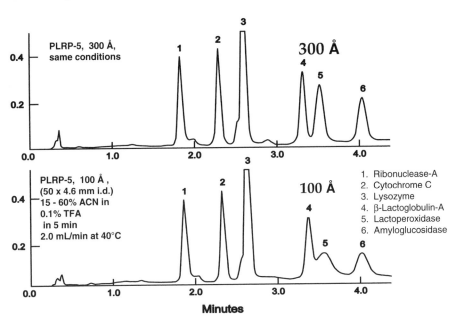

Figure 7.25. Comparative chromatograms illustrating the effect of pore size on protein separation. Two larger proteins (peaks 5 and 6) show substantial peak broadening due to entangled diffusion within the smaller pores of the second column. Reprinted with permission from reference 35.

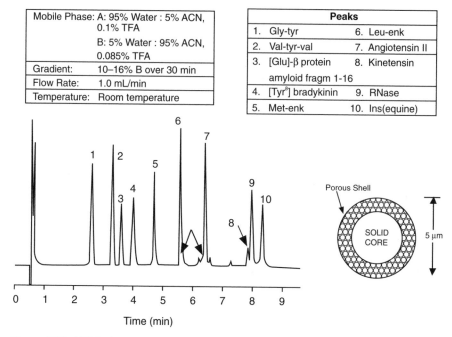

Mobile Phase:	A: 95% Water : 5% ACN, 0.1% TFA
	B: 5% Water : 95% ACN, 0.085% TFA
Gradient:	10–16% B over 30 min
Flow Rate:	1.0 mL/min
Temperature:	Room temperature

Peaks	
1. Gly-tyr	6. Leu-enk
2. Val-tyr-val	7. Angiotensin II
3. [Glu]-β protein amyloid fragm 1-16	8. Kinetensin
4. [Tyr⁸] bradykinin	9. RNase
5. Met-enk	10. Ins(equine)

Figure 7.26. HPLC chromatogram showing the performance of a pellicular column (Zorbax Poroshell 300-SB C4, 50 × 4.6 mm, 5 μm) in fast protein separation. Inset shows the structure of Zorbax Poroshell 300-SB particles. Diagram courtesy of Agilent Technologies.

>300 Å material. Note that wide-pore packings typically have lower surface area and therefore lower sample capacities. Other approaches for bioseparations are the use of very-wide-pore materials,[34,45] nonporous,[35] or pellicular materials (Figure 7.26). Nonporous and pellicular materials are particularly useful for fast separations.

Common protein purification strategies from natural sources or fermentation broths might involve a sequence of separation modes such as IEC–RPC–SEC, IEC–HIC (Affinity) / RPC,[33,34] etc. An example of IEC of large proteins using a weak anion exchange resin column eluted with a gradient of aqueous buffer with increasing salt concentration is shown in Figure 7.27. An example of SEC of large proteins, often termed gel-filtration chromatography (GFC), is shown in Figure 7.28. GFC is a gentle technique useful for the separation of protein conformers or aggregates.[34]

In contrast to the chromatography of large proteins, peptides are separated effectively with columns packed with more conventional C18-bonded phases on small-pore or medium-pore materials.[35] Peptide mapping is a common protein characterization technique in which a protein is cleaved by an enzyme to yield peptide fragments, which are then separated by HPLC to yield a peptide map[36] (Figure 7.29). Prior to the 1990s, these peptide fragments would be collected for Edman sequencing to yield the amino acid sequence for the protein. This tedious process of performing Edman sequencing on collected

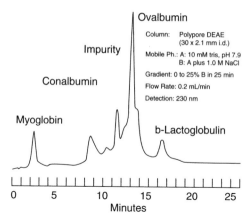

Figure 7.27. HPLC analysis of a protein mixture using weak anion-exchange chromatography using a polymer-based column and UV detection at 230 nm. Chromatogram courtesy of PerkinElmer, Inc.

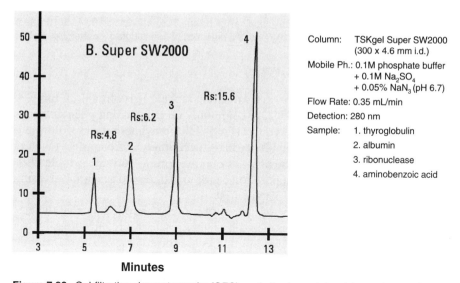

Minutes

Figure 7.28. Gel-filtration chromatography (GFC) analysis of a protein mixture using a polymer-based column and UV detection at 280 nm. Chromatogram courtesy of TOSOH Corporation.

peptide fragments is now mostly superseded by the use of LC/MS to yield sequence information directly during peptide mapping. An ion trap or hybrid quadrupole-time-of-flight (Q-TOF) MS with higher mass accuracy coupled to an electrospray interface is most suitable for this *de novo* sequencing using mass spectrometry.[32]

Amino acid analysis has been discussed in the food applications section of this chapter. In life science applications where sample size can be very limited,

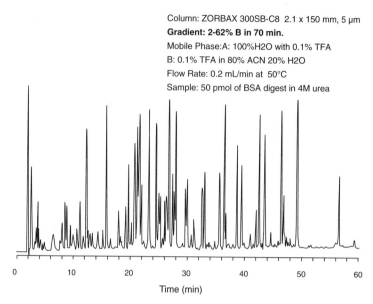

Column: ZORBAX 300SB-C8 2.1 x 150 mm, 5 µm
Gradient: 2-62% B in 70 min.
Mobile Phase:A: 100%H2O with 0.1% TFA
B: 0.1% TFA in 80% ACN 20% H2O
Flow Rate: 0.2 mL/min at 50°C
Sample: 50 pmol of BSA digest in 4M urea

Figure 7.29. A tryptic map (peptide mapping) of bovine serum albumin (BSA, 50 pmoles) using reversed-phase chromatography and UV detection. Chromatogram courtesy of Agilent Technologies.

the high-sensitivity pre-column derivatization technique is often preferable. A number of derivatization agents are commonly used, including phenylisothio-cyanate (PITC), o-phthaldehyde (OPA), 9-fluorenylmethyl chloroformate (FMOC), and 6-aminoquinolyl-N-hydroxysuccinimidyl carbamate[37] (Waters AccQ.Tag). An example on the analysis of a protein hydrolyzate sample illustrating the performance and separation conditions of amino acid analysis using Waters AccQ.Tag is shown in Figure 7.30.

Proteomics, the study of the entire set of proteins encoded by a genome, is an area of active research conducted by many research organizations.[32–38] As mentioned in Chapter 4, proteomics samples are too complex to be sufficiently resolved by a single HPLC column with a typical peak capacity of 200–400. However, multi-dimensional chromatography with two orthogonal columns can potentially extend peak capacity by ~15,000. The traditional approach is to use IEC (strong cationic, SCX) to fractionate the complex sample, followed by RPC-MS/MS to characterize each fraction, as shown in the example in Figure 7.31.

7.6.2   Bases, Nucleosides, Nucleotides, Oligonucleotides, Nucleic Acids, and PCR Products

HPLC analysis of bases and small nucleotides are best accomplished by RPC or IEC,[31,39] as exemplified by the example shown in Figure 7.32.

Sample: AQC-Derivatized Amino Acid Standards
Column: AccQ•Tag™ Column, 3.9 mm x 150 mm
Temperature: 37°C
Eluent A: AccQ•Tag™ Eluent A
Eluent B: Acetonitrile
Eluent C: HPLC grade water
Flow Rate: 1 mL/min
Gradient: AccQ•Tag™ Method
Detection: Fluorescence: lex = 250 nm, lem = 395 nm

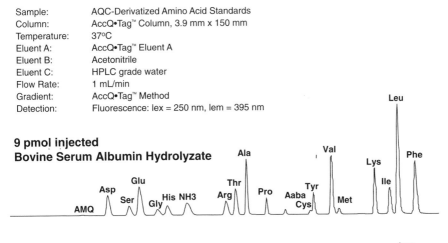

**9 pmol injected
Bovine Serum Albumin Hydrolyzate**

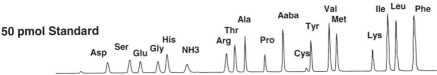

50 pmol Standard

Figure 7.30. HPLC analysis of pre-column derivatized amino acids (50 pmoles) using Waters AccQ.Tag reagents with reversed-phase chromatography separation and fluorescence detection. Chromatogram courtesy of Waters Corporation.

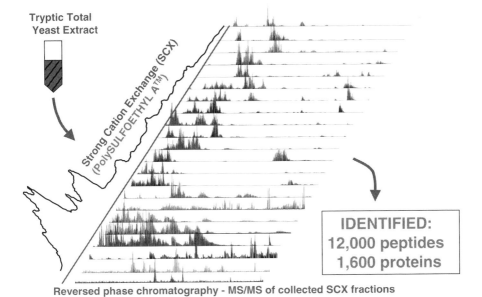

Figure 7.31. An example of a two-dimensional LC/MS analysis of yeast proteome using cation-exchange (SCX) for fractionation and reversed-phase LC/MS/MS of the collected fractions. Diagram courtesy of Andy Alpert of Poly LC from data originated by S. Gygi and Junmin Peng of Harvard Medical School.

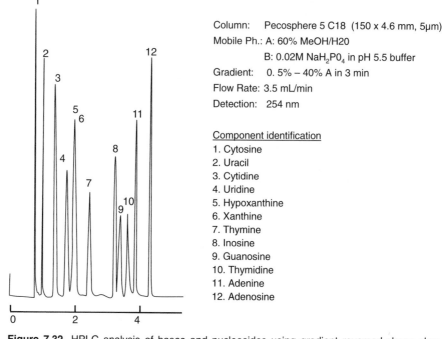

Column:	Pecosphere 5 C18 (150 x 4.6 mm, 5μm)
Mobile Ph.:	A: 60% MeOH/H20
	B: 0.02M NaH$_2$PO$_4$ in pH 5.5 buffer
Gradient:	0. 5% – 40% A in 3 min
Flow Rate:	3.5 mL/min
Detection:	254 nm

Component identification
1. Cytosine
2. Uracil
3. Cytidine
4. Uridine
5. Hypoxanthine
6. Xanthine
7. Thymine
8. Inosine
9. Guanosine
10. Thymidine
11. Adenine
12. Adenosine

Figure 7.32. HPLC analysis of bases and nucleosides using gradient reversed-phase chromatography and UV detection at 254 nm. Chromatogram courtesy of PerkinElmer, Inc.

Oligonucleotides, DNA restriction fragments, and PCR products are analyzed by gel electrophoresis, capillary electrophoresis, chip-based devices, or HPLC using IEC[34,40] (Figure 7.33) or ion-pair RPC[31,33,34] (Figure 7.34). HPLC is most useful for quantitation and purification of oligonucleotides and PCR products. Typical HPLC assay performance characteristics using a nonporous anion-exchange column are as follows: molecular size range (1–20,000 base pairs), sensitivity (0.2–0.4 ng), quantitation (10%), loading capacity (0.5–10 μg), recovery (85–100%), and analysis time (10–20 min).[40]

7.7 SUMMARY

HPLC is a versatile technique applicable to diversified analytes, including labile molecules, ions, organic, and biopolymers. This chapter provides an overview of HPLC applications for the analysis of food, environmental, chemical, polymer, ion-chromatography, and life science samples. In food analysis, HPLC is widely used in product research, quality control, nutritional labeling, and residual testing of contaminants. In environmental testing, HPLC is excellent for the sensitive and specific detection of labile and nonvolatile pollutants

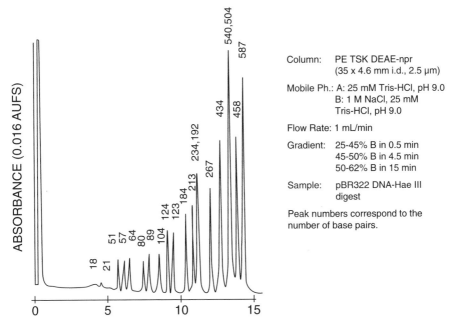

Figure 7.33. HPLC analysis of nuclei acids (pBR322-DNA HAE-III digest) using gradient weak-anion exchange chromatography on a column packed with 2.5-μm nonporous polymer support (weak anion exchange) with UV detection at 260 nm. Reprinted with permission from reference 40.

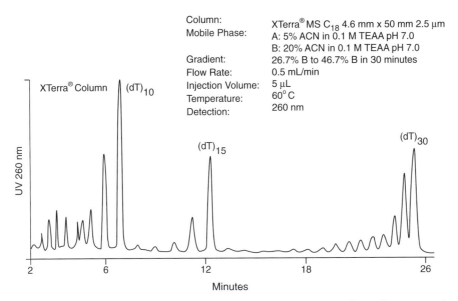

Figure 7.34. HPLC analysis of synthetic oligonucleotides using ion-pair gradient reversed-phase chromatography on a column packed with 2.5-μm nonporous hybrid particles with UV detection at 260 nm. Chromatogram courtesy of Waters Corporation.

in drinking water, wastewater, solid waste, and air samples. HPLC applications in the chemical and plastics industries include assays of synthetic chemicals and polymer characterization in product research and quality control. Finally, HPLC life science applications focus on the separation, quantitation, and purification of biomolecules in medical research and drug discovery.

7.8 REFERENCES

1. L. Nollet, ed., *Food Analysis by HPLC*, 2nd Edition, Marcel Dekker, New York, 2000.
2. R. McCrae, ed., *HPLC in Food Analysis*, Academic Press, London, 1988.
3. A. Grasfeld-Husgen and R. Schuster, *HPLC for Food Analysis: A Primer*, Hewlett-Packard (Agilent Technologies), Waldbronn, Germany, 1996, Publication number 12–5965–5124E.
4. R. Helrich, ed., *Official Methods of Analysis of the Association of Official Analytical Chemists (AOAC)*, 15th edition, Arlington, VA, 1990.
5. S. Chang, et al. Food analysis—applications review, *Anal. Chem.* **65**, 334R (1993).
6. Food Labeling Regulations: Final Rule, *Federal Register, Volume 58*, No. 3. P.7066, 6, 1993.
7. F. Scott, in L. Nollet, ed., *Food Analysis by HPLC*, Marcel Dekker, New York, 1992.
8. M.W. Dong, *HPLC System for Carbohydrate Analysis, Carbohydrate Analysis Cookbook*, Perkin-Elmer, Norwalk, CT, 1993.
9. D. Marini, in L. Nollet, ed., *Food Analysis by HPLC*, Marcel Dekker, New York, 1992.
10. H.E. Indyk, *Analyst,* **115**, 1525 (1990).
11. M.W. Dong and J.L. DiCesare, *J. Amer. Oil Chem. Soc.* **60**, 788 (1983).
12. D. Blanco Gomis, in L. Nollet, ed., *Food Analysis by HPLC*, Marcel Dekker, New York, 1992.
13. M.W. Dong, *LC.GC* **16(12)**, 1092 (1998).
14. J. Vervaeck and A. Huyghebaert, in L. Nollet, ed., *Food Analysis by HPLC*, Marcel Dekker, New York, 1992.
15. J. White, et al., in L. Nollet, ed., *Food Analysis by HPLC*, Marcel Dekker, New York, 1992.
16. K. Saag, in R. Macrae, ed., *HPLC in Food Analysis,* Academic Press, London, 1988.
17. M.W. Dong and J.L. Pace, *LC.GC* **14 (9)**, 794 (1996).
18. M.W. Dong, *Today's Chemist at Work*, **9(5)**, 17 (2000).
19. J. Leitao, et al., in L. Nollet, ed., *Food Analysis by HPLC*, Marcel Dekker, New York, 1992.
20. J.N.A. Botsoglou, in L. Nollet, ed., *Food Analysis by HPLC*, Marcel Dekker, New York, 1992.
21. R.J. Bushway, in L. Nollet, ed., *Food Analysis by HPLC*, Marcel Dekker, New York, 1992.

22. Z.A. Grosser, J.F. Ryan, and M.W. Dong, *J. Chromatog.* **642**, 75 (1993).

23. A. Grasfeld-Husgen and R. Schuster, *HPLC for Environmental Analysis: A Primer*, Hewlett-Packard (Agilent Technologies), Germany, 1997 Publication number 12-5091-9750E.

24. *HPLC in Environmental Testing*, Order Number L-1534, Perkin-Elmer, Norwalk, CT, 1992.

25. M.W. Dong, M.V. Pickering, M.J. Mattina, and H.M. Pylypiw, *LC.GC* **10**, 442 (1992).

26. M.W. Dong, J.X. Duggan, and S. Stefanou, *LC.GC* **11(11)**, 802 (1993).

27. S. Mori and H.G. Barth, *Size Exclusion Chromatography*, Springer-Verlag, Berlin, Germany, 1999.

28 W.M. Reuter, M.W. Dong, and J. McConville, *Amer. Lab.* **23(5)**, 45 (1991).

29. J. Weiss, *Ion Chromatography*, 2nd Edition, VCH, Weinheim, Germany, 1995.

30. C. Pohl, in S. Ahuja and M.W. Dong, eds., *Handbook of Pharmaceutical Analysis by HPLC*, Elsevier, Amsterdam, 2005.

31. M.A. Vijayalakshmi, *Biochromatography: Theory and Practice*, Taylor & Francis, London, 2002.

32. J.-C. Sanchez, G.L. Corthais, and D.F. Hochstrasser, *Biomedical Applications of Proteomics*, John Wiley & Sons, Hoboken, NJ, 2002.

33. *Waters Life Sciences Chemistry Solutions Brochure*, Waters Corp., Milford, MA 2004, Library No. 720000831EN.

34. R.L. Cunico, K. M. Gooding, and T. Wehr, *Basic HPLC and CE of Biomolecules*, Bay Bioanalytical Laboratory, Richmond, CA, 1998.

35. M.W. Dong, J.R. Gant, and B. Larsen, *BioChromatogr.* **4(1)**, 19 (1989).

36. M.W. Dong, in P. Brown, ed., *Advances in Chromatography*, Volume. 32, Marcel Dekker, New York, 1992, pp. 21–51.

37. D. Strydom and S. Cohen, *Anal. Biochem.*, **222**, 19 (1994).

38. J. Masuda et al., *J. Chromatogr.* A, **1063**, 57 (2005).

39. C.W. Gehrke and K.C. Kuo, in C.W. Gehrke and K.C. Kuo, eds., *Chromatography and Modifications of Nucleosides*, Elsevier Chromatography Library Series, Volume 45A, Amsterdam, The Netherlands, 1990, pp. A3–A7.

40. E. Katz and M.W. Dong, *BioTechniques,* **8(5)**, 546 (1990).

8

HPLC METHOD DEVELOPMENT

Modern HPLC for Practicing Scientists, by Michael W. Dong
Copyright © 2006 John Wiley & Sons, Inc.

8.1   INTRODUCTION

8.1.1   Scope

This chapter provides an overview of modern HPLC method development and discusses approaches for initial method development (column, detector, and mobile phase selection), method optimization to improve resolution, and emerging method development trends. The focus is on reversed-phase methods for quantitative analysis of small organic molecules since RPLC accounts for 60–80% of these applications. Several case studies on pharmaceutical impurity testing are presented to illustrate the method development process. For a detailed treatment of this subject and examples of other sample types, the reader is referred to the classic book on general HPLC method development by L. Snyder et al.[1] and book chapters[2, 3] on pharmaceutical method development by H. Rasmussen et al. Other resources include computer-based training[4] and training courses.[5]

8.1.2   Considerations Before Method Development

Developing and validating new analytical methods is costly and time consuming. Before starting the arduous process, a thorough literature search should be conducted for existing methodologies of the intended analytes or similar compounds. This should include a computerized search of chemical abstracts and other relevant sources such as compendial monographs (USP, EP), journal articles (*JAOAC, ASTM, J. Chromatography*), manufacturer literature, and the Internet. Although this search might not uncover a directly usable method, it often provides a starting point for method development or at least some useful references.

New analytical methods are needed for the following reasons:

- Existing methods are not available (e.g., New Chemical Entity (NCE) for consideration as a new drug candidate).
- Existing methods are not sufficiently reliable, sensitive, or cost effective.
- New instrumentation or technique has better performance (ease of use, rapid turnaround, automation, higher sensitivity).
- An alternate (orthogonal) method is required for regulatory compliance.

8.1.3 Strategy for Method Development

Steps in a common strategy for HPLC method development are summarized below:

1. Define method and separation goals
2. Gather sample and analyte information
3. Initial method development — "scouting" runs and getting the first chromatograms
4. Method fine-tuning and optimization
5. Method validation

Steps 1–4 are discussed here in Sections 8.2–8.5, while method validation is discussed in Chapter 9, Section 9.3.

8.1.4 HPLC Method Development Trends in Pharmaceutical Analysis

Some of the "best practices" and emergent trends in pharmaceutical method development are highlighted here and discussed further as case studies:

- Modern instrumental trends for method development include:
  - Multi-solvent pumps, photodiode array detectors (PDA), modeling software, automated development systems, and multi-column selector valves
- Trends toward MS-compatible, gradient methods for impurity testing
- Composite methods (combined assay/impurity) during early development
- Phase-appropriate method development and validation
- Use of secondary orthogonal methods to ensure separation of all impurities and unexpected unknowns
- A single method for products of all different strengths and formulations
- LC/MS/MS methods for trace analysis (e.g., bioanalytical assays, trace genotoxic impurities)

Although most pharmaceutical methods use UV detection, the importance of adopting MS-compatible mobile phases cannot be overemphasized. MS is invaluable for quick identification of unknowns often encountered during early process development and stability studies. The use of composite (combined assay and impurity) methods to facilitate testing is increasing. Area normalization is often used in early-stage methods due to a lack of impurity reference standards. The concept of phase-appropriate method development is receiving more attention as described further in Section 8.6 and in recent publications.[2, 3]

8.2   DEFINING METHOD TYPES AND GOALS

HPLC methods are developed for single-analyte or for multiple-analyte assays. Methods can be categorized into three major method types: qualitative, quantitative, and preparative.

A *qualitative method* is primarily an identification test that confirms the presence (or absence) of a certain analyte(s) in the sample by matching retention time with that of a reference standard. UV spectral data from a photodiode array detector are often used as a secondary confirmation technique. This type of method can be a limit test to evaluate whether the level of the analyte is above or below a certain preset limit or to generate a chromatographic profile for comparative purposes.

A *quantitative method* generates information on the concentration or amount of the analyte(s) in the sample. System calibration (standardization) typically using external standard(s) is required. A quantitative method can also be adopted as a qualitative method. For instance, an assay method can often also serve as an identification method. A quantitative method is more difficult to develop and requires extensive effort for validation. This method type is the focus of this chapter.

A *preparative method* is used to isolate purified components in the sample. This method often requires large-diameter columns and high flow rates to increase yield for each run. Method validation is typically not required since the goal is to generate purified components or enriched fractions.

8.2.1   Method Goals

Analytical method goals are often defined as method acceptance criteria for peak resolution, precision, specificity, and sensitivity. For instance, pharmaceutical methods for potency assays of an active pharmaceutical ingredient (API) typically require the following: resolution ≥1.5 from the closest eluting components; precision of retention time and peak area, <1–2% RSD; and linearity in the range of 50–150% of the label claim. Other desirable characteristics include:

- Analysis time ~5–30 min (<60 min for complex samples)
- Minimal sample work-up (extract and inject if possible)

Table 8.1. Pertinent Sample and Analyte Information

Sample/analyte	Information
Sample	Number of components
	Concentration range of analytes
Analyte(s)	Chemical structure and molecular weight
	pK_a
	Solubility
	Chromophore, maximum absorbance wavelength (λ_{max})
	Chiral centers, isomers
	Stability and toxicity
Others	Availability and purity of reference standard materials

- Robust method that does not require extensive training for execution
- Low cost per analysis (low reagent, instrument, and set-up cost)

8.3 GATHERING SAMPLE AND ANALYTE INFORMATION

After defining method goals, the next step is to gather sample and analyte(s) information such as those listed in Table 8.1. This information is useful for the selection of appropriate sample preparation procedures as well as the initial detection and chromatographic modes. If critical data are not available (e.g., pK_a, solubility), separate studies should be initiated as soon as possible.

The chemical structure of the analyte furnishes data on molecular weight and the nature of the functional groups. Particular attention should be directed to acidic, basic, aromatic, or reactive functional groups from which estimates of pK_a, solubility, chromophoric, or stability data can be inferred. If sufficient purified reference material is available, solubility studies of the analyte in common solvents such as water, alcohol, ether, and hexane should be conducted. Toxicity data and Material Safety Data Sheets (MSDS) should be gathered if available to develop appropriate guidelines on safe handling procedures. The availability and the purity of any reference standard materials should be ascertained. Certificates of analysis (COA) from the material vendors can be an invaluable source of pertinent data, including reference chromatograms and spectral data (MS, NMR, IR, and UV).

8.3.1 Defining Sample Preparation Requirements

Table 8.2 lists common sample preparation techniques for extraction and enrichment of analytes before HPLC analysis. More detailed discussion on sample preparation procedures is available elsewhere.[6, 7] A modern method trend is the use of higher-resolution separation techniques coupled with specific detectors (e.g., LC/MS/MS) to eliminate or minimize the time-consuming

Table 8.2. Common Sample Preparation Procedures

Type	Operation	Comments
Solid handling	Grinding, milling, homogenization	To reduce particle size of sample to facilitate extraction
Extraction	Shaking, ultrasonication, Soxhlet, liquid-liquid partitioning, solid-phase extraction	To extract analytes from sample matrix into solution
Liquid handling	Pipetting, dilution, concentration, pH/ionic strength adjustment	To adjust the volume or attribute of the analyte extract
Phase separation	Filtration, centrifugation, precipitation	To separate analytes from other sample matrices
Sample clean-up	Column chromatography, HPLC, GC, TLC, GPC	To fractionate and enrich analytes from complex samples matrices
Derivatization	Pre-column of post-column chemical derivatization	To transform the analyte into a form to improve sensitivity or selectivity

sample preparation steps. This trend is most evident in bioanalytical testing of drugs and their metabolites in physiological fluids samples (serum, plasma), which is mostly performed by LC/MS/MS techniques.

8.4   INITIAL HPLC METHOD DEVELOPMENT

During initial method development, a set of initial conditions (detector, column, mobile phase) is selected to obtain the first "scouting" chromatograms of the sample. In most cases, these are based on reversed-phase separations on a C18 column with UV detection. A decision on developing either an isocratic or a gradient method should be made at this point.

8.4.1   Initial Detector Selection

For analytes having reasonable UV absorbance, the UV/Vis detector (or the PDA detector) is the clear choice. As discussed in Chapter 4, UV/Vis detectors are reliable, sensitive, easy-to-use, and very precise. The actual monitoring wavelength is usually set at the maximum absorbance wavelength of the analyte or at a far UV wavelength (i.e., 200–230 nm) to improve sensitivity (Figure 8.1a–c). For nonchromophoric analytes (with low or no absorbance in UV-Vis range), the choice is limited to refractive index (RI) detector or evaporative light scattering detector (ELSD) for gradient analysis. Mass spectrometry (MS) is a possible choice for "ionizable" analytes. Although MS (or

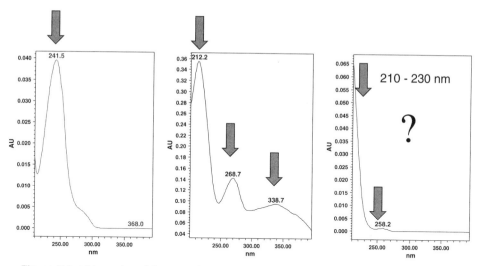

Figure 8.1. UV spectra of three analytes illustrating the decision process in setting HPLC monitoring wavelength. (A) The λ_{max} at 241 nm is the clear choice. (B) The three λ_{max} at 212 nm, 269 nm, and 339 nm give rise to different possibilities: 269 nm is the obvious choice though 212 nm can be selected for higher sensitivity or 339 nm for higher selectivity if interference from other matrix components is a problem. (C) The λ_{max} at 258 nm is a clear choice if sensitivity is not an issue, otherwise the wavelength at far UV (210–230 nm) are better selection for higher sensitivity. The actual choice in the far UV range is often dependent on baseline gradient shifts due to the absorbance of the mobile phase at that the monitoring wavelength.

MS/MS) is the standard detector for bioanalytical assays and drug discovery screening, its use for routine assays of drug substances and products is still limited due to its high cost and lower precision. Nevertheless, LC/MS/MS methods are increasingly used for ultra trace analysis or screening of complex samples. Other detection options include conductivity detection for ionic species and electrochemical detection for neuroactive species in biochemical research.

8.4.2 Selection of Chromatographic Mode

Table 8.3 lists some common chromatographic modes based on the analyte's molecular weight and polarity. All case studies will focus on reversed-phase chromatography (RPC), the most common mode for small organic molecules. Note that ionizable compounds (acids and bases) are often separated by RPC with buffered mobile phases (to keep the analytes in a non-ionized state) or with ion-pairing reagents.

8.4.3 Initial Selection of HPLC Column

Chapter 3 discusses the characteristics of columns and packing materials. Although there is a wide choice of RPC-bonded phases on the market, a silica-

Table 8.3. Guideline for Chromatographic Mode Selection

Sample type	Analytes type	Common mode
Macromolecules (MW > 2,000)	Organic polymers	GPC
	Biomolecules	SEC, RPC, IEC, HILIC, HIC
Organics (MW < 2,000)	Polar	RPC, NP, HILIC
	Medium polarity	RPC
	Nonpolar	RPC, NARP, NP
	Ions, ionizable compounds	RPC (ion suppression), RPC-IP, IEC, HILIC
Preparative	All	NP, RPC, GPC, IEC

Reversed-phase (RPC), ion-pair (IP), ion-exchange (IEC), hydrophilic interaction (HILIC), hydrophobic interaction (HIC), normal-phase (NP), gel permeation (GPC), size-exclusion (SEC), nonaqueous RP (NARP).

based C18 or C8 column remains a good starting point because of its high efficiency and stability. Additional guidelines for initial column selection follow. Select columns packed with 3- or 5-µm high-purity silica-bonded phases from a reputable manufacturer. Three-micrometer packing is probably preferable due to its faster analysis since shorter column length can be used. However, columns packed with sub-2-µm packing should be purchased with caution since they are best used for clean samples on low-dispersion instruments. The following column dimensions are suggested:

- 50–100 mm × 4.6 mm i.d. for simple samples (e.g., assays of the main component)
- 100–150 mm × 3.0–4.6 mm i.d. for purity testing or multi-component testing of complex samples
- 20–150 mm × 2.0 mm columns for LC/MS

Guidelines on mobile phase selection are discussed in Chapter 2, Section 2.3 and are illustrated in the case study below. After evaluation of the first sample chromatograms, further method development to fine-tune the separation will be performed until all method goals are achieved. Other bonded phases or column configurations can be selected to enhance method performance. These are discussed in the Section 8.6.

8.4.4 Generating a First Chromatogram

8.4.4.1 Case Study: Initial Method Development Using a Broad Gradient and Mobile Phase Selection
In this case study, a broad scouting gradient is used to obtain the first sample chromatogram. The method goal is to develop an MS-compatible composite method (combined assay and impurity) for a drug substance. This pharmaceutical active ingredient (API) is a basic

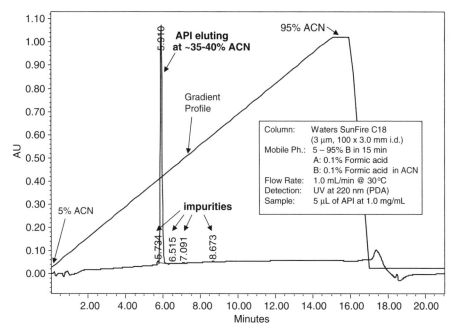

Figure 8.2. The first HPLC chromatogram of a drug substance analyzed with generic broad gradient conditions. The active pharmaceutical ingredient (API) elutes at %B of 35–40% of acetonitrile, showing presence of several impurities between 5 and 9 min.

salt with a pK_a of ~8. It has good solubility in water, especially under acidic pH, and has a UV spectrum shown in Figure 8.1C. A high-purity silica-based C18 column (3 μm, 100 × 3.0 mm) was selected to yield a first chromatogram with a broad gradient (mobile phase A, MPA = 0.1% formic acid in water; mobile phase B, MPB = 0.1% formic acid in acetonitrile), as shown in Figure 8.2. An acidic mobile phase was chosen as it typically yields better peak shapes. A photodiode array detector (PDA) was used to collect data between 210 and 400 nm, allowing chromatograms to be plotted at 260 nm, 230 nm, 220 nm, and 210 nm, as shown in Figure 8.3. As expected from the UV spectrum of the compound, far UV detection at 210–230 nm yielded more sensitive detection, though significant baseline gradient shifts were discernable. At this point, data suggested that the optimum detection wavelength be 220 nm for this mobile phase since it produces the least gradient shift of the baseline.

Figure 8.2 shows all sample peaks elute between 5 and 9 min (or 30–50% ACN). The gradient range was thus reduced to 20–60% ACN yielding a chromatogram with better resolution of the impurities (Figure 8.4). This is our initial gradient method. Note that method sensitivity is enhanced for impurities by injecting sufficient sample to yield an absorbance of ~1.0 AU for the API, below the linearity limits of this PDA detector.

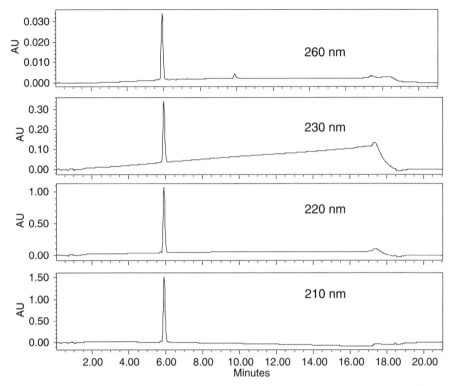

Figure 8.3. The same chromatographic data of Figure 8.2 were plotted at 260 nm, 230 nm, 220 nm, and 210 nm from the archived PDA data. As expected from the UV spectrum of the compound, far UV detection at 210–230 nm yielded more sensitive detection, though some baseline gradient shifts were also discernable. Two hundred and twenty nanometers appears to be an excellent monitoring wavelength yielding an absorbance of the API at ~1 absorbance unit (AU) accompanied by a relatively flat baseline.

If an isocratic assay method is needed (i.e., for assay or dissolution testing), the optimum mobile phase conditions can be conveniently found by using an approach termed "sequential isocratic steps." This entails injecting the sample using successively lower percentages of organic solvent (ACN) in the mobile phase, as shown in Figure 8.5, until optimal chromatography is achieved. The best condition for isocratic assay for this analysis was found at 25% CAN, as shown in Figure 8.6.

This case study illustrates the logical sequence of using a generic scouting gradient for an overall sample assessment and establishing the initial chromatographic conditions. If these scouting runs fail to establish adequate chromatograms (e.g., insufficient retention, bad peak shape, or poor sensitivity), then other approaches should be explored (e.g., use of ion-pairing reagents, other chromatographic modes, or alternate columns).

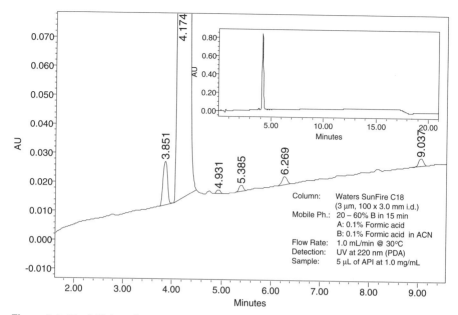

Figure 8.4. The initial gradient method for this API was obtained by reducing the gradient range to 20–60% ACN, yielding better resolution of the impurities. The inset is the full-scale chromatogram showing the absorbance of the API peak to be ~1 AU.

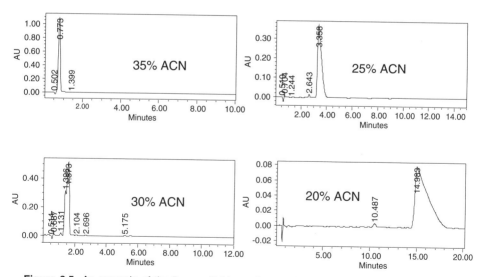

Figure 8.5. An example of the "sequential isocratic steps" approach by injecting the sample with successively lower percentage of mobile phase B (ACN) used to find the optimum isocratic mobile phase. Note that peak area becomes severely tailing at 20% ACN due possibly to solubility issues in the weaker mobile phase.

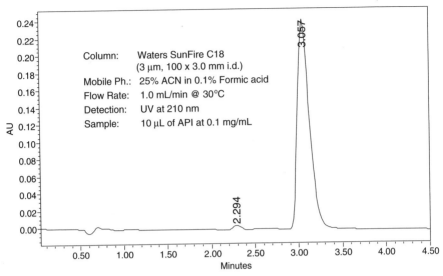

Figure 8.6. The optimum mobile phase for isocratic assay is at found at 25% ACN.

8.5 METHOD FINE-TUNING

The basic goal of most chromatographic methods is to achieve adequate resolution of all key analytes with sufficient precision and sensitivity in a reasonable time. After obtaining the initial separation conditions, the next step is
to fine-tune each parameter to meet all requisite goals. Often, the first step is
to confirm method "specificity" — to assure that all analyte peaks are resolved
from other potential interfering components in the sample (see Chapter 9).
This is accomplished by confirming noninterference from the procedural blank
(diluent) or placebo sample (sample containing all formulation components
except the active ingredient). For impurity testing, a test mixture or "cocktail"
of process impurities and degradants is used to evaluate the separation capability of the initial method. If these impurity reference standards are not available, a "stressed" or force-degraded sample is used to challenge the method
(see Chapter 9, Section 9.3). Much effort is expended to make sure impurities
are not "hidden" under the main API peak. This is usually done by evaluating
peak purity by PDA or MS, and by developing an "orthogonal" method using
a different separation mechanism than that of the primary method (see
Chapter 2).

   At this stage, any co-elution of the key analytes is corrected by fine-tuning
the separation parameters, mostly by adjusting selectivity (α) (see Figure 2.17).
This process may be quite challenging and iterative for multi-component
assays of complex samples. Afterwards, efforts are focused on improving other
method performance measurements such as sensitivity, peak shape, robust-

ness, and analysis time. This process is analogous to method validation except that the method is not "frozen" at this stage. An experienced method developer often spends more time at this stage to lessen problems during subsequent method validation. Parameters adjusted in this stage are:

- Mobile phase parameters
  — Percentage organic solvent (%B), buffer type and concentration, pH, solvent type
- Operating parameters
  — Flow (F), temperature (T), gradient range ($\Delta\phi$), gradient time (t_G)
- Column
  — Bonded phase type, length (L), column diameter (d_c), particle size (d_p)
- Detector setting (monitoring wavelength) and sample amount

8.5.1 Mobile Phase Parameters (%B, Buffer, pH, Solvent Type)

In HPLC, the mobile phase controls the retention and selectivity of the separation and can be conveniently and continuously adjusted during method improvement. Lowering the solvent strength (%B of the strong solvent) increases retention and typically increases resolution (see Figure 2.16). In isocratic separations, the analyte peaks are typically kept in the k range of 2–20 to maintain sufficient sensitivity of the late eluting peaks. Lowering solvent strength might result in peak shape distortion (Figure 8.5, from 25% to 20% ACN) caused by poor solubility of the analytes in the lower-strength mobile phases. Reducing %B lower than 3–5% B can also be problematic due to a phenomenon of phase collapse or "dewetting" the hydrophobic bonded groups of the stationary phase (see Figure 4.16). Common strategies for increasing retention and resolution of very polar compounds include ion pairing, or the use of polar-embedded or lower-coverage bonded phases (see Chapter 4).

Buffers in the mobile phase are required for controlling the separations of acidic or basic analytes (see Chapter 3, Section 3.3.3). Buffer concentrations of 10–25 mM levels are usually sufficient. Volatile buffers (formate, acetate, carbonate, bicarbonate, ammonia) are needed for MS-compatible methods. One major disadvantage for the use these volatile buffers (vs. phosphate) is a loss of sensitivity at far UV due to lack of transparency at these wavelength. Trifluroacetic acid (TFA) and heptafluorobutyric acid (HFBA) are excellent volatile acidifiers that also have some ion-pairing properties. pH is a powerful factor for fine-tuning resolution, as shown in Figures 2.13 and 2.14. The introduction of high-pH compatible silica-based bonded phases is a significant development.[8] Examples of high-pH separations are described in the Case Study in Section 8.8.3. Organic modifiers in RPC (MeOH, CAN, and THF) are traditionally used to adjust selectivity (Figure 2.17). Since THF is unstable and toxic, some analysts prefer the use of small amounts of methyl-t-butyl ether (MTBE) instead.

8.5.2 Operating Parameters (F, T, Δφ, t$_G$)

Flow rate (F) does not affect isocratic retention (k) or selectivity but is an important factor affecting both average retention (k*) and selectivity (α) in gradient elution (Figure 2.22). Increasing column temperatures (T) generally reduces retention in RPC and can have some effects on selectivity. Δφ is the range of percentage of the strong solvent (%B) in gradient analysis (final %B minus initial %B) and is typically defined during initial method development. t$_G$ is gradient time, which in conjunction with Δφ defines the gradient slope. Increasing t$_G$ often increases overall resolution of complex samples. Since both T and t$_G$ can be continuously varied, their combined change offers an exceptional opportunity for fine-tuning separations of complex samples as shown case studies A and C. The use of software simulation programs such as DryLab is highly recommended for efficient gradient optimization, since the predictions on the effects of gradient factors are not always intuitive.

8.5.3 Column Parameters (Bonded Phase Type, L, d$_p$, d$_c$,)

Selectivity of different bonded phases is a powerful factor in method development[1,2,8] (Figures 3.8 and 3.15). However, since column switching is not a continuous variable process like fine-tuning mobile phase conditions, it is often the last step to be implemented, when mobile phase adjustments do not yield the requisite resolution. One should pick a column with different selectivity (orthogonal) from the original column (Figure 2.24). Traditionally, C8, cyano, and phenyl give the most dissimilar selectivity (Figure 3.8). Recently developed polar-embedded phases are also excellent choices (Figure 3.15). As it is difficult to predict which one column will yield the best separation of a given mixture, the practical approach is to try different columns under the optimized mobile phase conditions or to optimize the mobile phase for the most promising candidate column. A six-column automatic column selector can be used to facilitate this column selection process. Column dimension (L, d$_p$, d$_c$) can also be optimized to enhance efficiency, speed, or sensitivity. The use of software simulation is particularly useful to find the optimum column dimension for a separation, as shown in the case study in Section 8.8.1.

8.5.4 Detector Setting and Sample Amount

Finally, detector setting and sample loading (sample concentration and injection volume) are finalized. The goal is to maximize sensitivity (increase signal-to-noise) while maintaining method linearity and peak shapes. Sensitivity is important for impurity methods and for trace analysis. For large-volume injections (i.e., >20 μL), the sample solvent strength should be equal to or weaker than the starting mobile phase to prevent chromatographic anomalies (Figure 10.11). Obviously, the maximum sample amount is highly dependent on column size.

8.5.5   Summary of Method Development Steps

Table 8.4 summarizes the factors and suggested guidelines in method development. Although each method development scientist might favor his or her own individual approach, the adjustment factors and sequence of steps are likely to be similar.

Table 8.4. Summary of Factors Optimized During RPLC Method Development

Stage/approaches	Factor	Suggestions and comments
Initial development		
Sequential isocratic steps or scouting gradients	Column	Modern columns packed with C8 or C18 phases of 3- or 5-μm high-purity-silica support
	Detection wavelength	Use PDA to evaluate λ_{max} of all analytes. Select λ_{max} of main component or far-UV wavelength (200–230 nm).
	Mobile phase	MPA: Water for neutral analytes or acidified at pH 2–4 for acidic or basic analytes
		MPB: MeOH or ACN
	Operating	Use broad gradient for initial evaluation
	Conditions	Evaluate first chromatogram to define gradient range $\Delta\phi$
		Set T = 30°C, F is defined by d_c
Method development and optimization		
Isocratic or gradient methods	Mobile phase and operating conditions	Evaluate method with impurity "cocktail" and placebo
		Modify solvent type, pH, T, and t_G to improve resolution
Method fine-tuning and finalization		
Isocratic or gradient methods	Column	Evaluate method with different bonded phases
		Optimize column dimension (DryLab recommended)
	Detector and sample	Finalize detection wavelength and sample concentration or injection volume

8.6 PHASE-APPROPRIATE METHOD DEVELOPMENT

Phase-appropriate method development, is a proactive method development approach advocated by H. Rasmussen et al.[2,3] to correlate the method development process with the requirements of the clinical and regulatory phases during pharmaceutical development (Figure 8.7). Level 1 or level 2 composite methods tend to be broad-gradient, MS-compatible methods developed and validated for early-phase development. These primary methods are backed up by a secondary (orthogonal) methods used to evaluate pivotal lots of drug substance and early formulations. Examples of primary and secondary methods are shown in Figure 8.8. The secondary method can be quickly validated if the primary method is found deficient. Level 3 methods are developed after the filing of a New Drug Application (NDA) when synthetic routes, formulation, processing conditions, and specifications are "locked." Its purpose is to lead to quick, robust, and transferable final Level 4 methods, which will undergo more thorough validation. Level 4 methods are often isocratic methods yielding robust quantitation of the specified degradants and impurities. Some examples of the columns and mobile phase conditions suggested by Rasmussen et al.[2] for the development of early-phase methods are shown in Table 8.5.

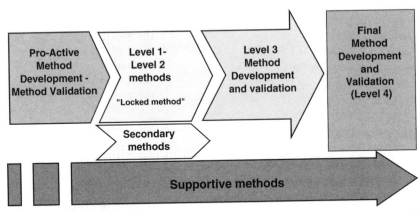

Development level	Level 1	Level 2	Level 3/Level 4
Clinical phase	Phase 1	Phase 2	Phase 3
Regulatory phase	CTA/SAD/IND	IND/CTA	CTA/NDA/MAA/PAS

CTA:	Clinical Trial Application	NDA:	New Drug Application
IND:	Investigational New Drug	PAS:	Post Approval Support
MAA:	Marketing Authorization Application	SAD:	Safety Assessment Document

Figure 8.7. A schematic diagram showing the "phase-appropriate method development and validation approach" of Rasmussen et al. Reprinted with permission from reference 2.

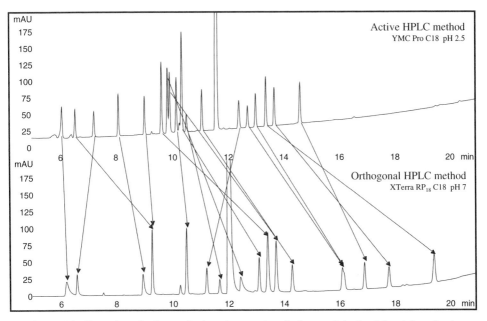

Figure 8.8. Examples of primary (top) and secondary (orthogonal, bottom) level 1 methods. Reprinted with permission from reference 2.

Table 8.5. Suggested Columns and Mobile Phase Conditions for Early-Phase Methods

HPLC columns ($100 \times 4.6\,mm$, $3\,\mu m$)
Phenomenex Luna phenyl-hexyl
Agilent Zorbax Bonus RP and Extend C18
YMC Pro C 18
Waters Symmetry-Shield RP 18, XTerra RP and MS
Thermo Hypersil BDS C 18

Mobile phase conditions (broad-gradient LC/UV and LC/MS methods)

Method A	MPA: $10\,mM$ NH_4OAc in water—CH_3CN (95/5) + 0.1% TFA
[pH = 2.5]	MPB: $10\,mM$ NH_4OAc in water—CH_3CN (15/85 + 0.1% TFA
Method B	MPA: $10\,mM$ NH_4OAc in water—CH_3CN (95/5) + 0.05% HOAc
[pH = 4.8]	MPB: $10\,mM$ NH_4OAc in water—CH_3CN (15/85) + 0.05% HOAc
Method C	MPA: $10\,mM$ NH4OAc in water—CH_3CN (95/5)
[pH = 7]	MPB: $10\,mM$ NH4OAc in water—CH_3CN (15/85)
Method D	MPA: $10\,mM$ $(NH_4)_2CO_3$ in water—CH_3CN (95/5)
[pH = 9]	MPB: $10\,mM$ $(NH_4)_2CO_3$ in water—CH_3CN (15/85)

MPA and MPB: mobile phase A and B. Table reprinted with permission from reference 2.

8.7   METHOD DEVELOPMENT SOFTWARE TOOLS

Table 8.6 provides descriptions and estimated costs for a number of software and automated systems for HPLC method development and optimization. The reader is referred to some key references[10-12] and the vendors' websites for further details. Examples on DryLab simulation, the first and the mostly widely used software, are shown in the case studies (Sections 8.8.1–8.8.3) of this chapter.

8.8   CASE STUDIES

Three method development cases studies of pharmaceuticals are used to illustrate the logical sequence in fine-tuning conditions to meet method goals. These are:

- A drug substance method for a neutral molecule illustrating the use of DryLab software for mobile phase and column optimization
- A drug substance method for a basic molecule illustrating the selection of bonded phases and mobile phases to improve resolution and sensitivity
- An impurity method for a drug product with two APIs illustrating the use of high-pH mobile phases and the optimization of pH, organic modifier and column

8.8.1   Composite Assay Method for a Neutral Drug Substance

This case study started with an initial isocratic method used for the evaluation of a new drug substance shown in Figure 8.9. Note that a mobile phase buffer was not used since the molecule is neutral. This separation was judged inadequate since two impurities (imp 4 and 4a) were not separated. Initial gradient assessment also revealed the presence of a late eluting impurity (imp 5). Since the number of impurities is limited, DryLab was applied immediately after the initial scouting gradient run to explore the optimum temperature (T) and gradient conditions. Four gradient runs were performed with the same column (Waters Symmetry C8, 150×3.9 mm, 5μm) under a broad gradient (35–100% MeOH) at two different temperatures (T = 23 and 50°C) and two gradient times (t_G = 30 and 90 min). Retention data of all peaks were entered manually into the DryLab software to allow simulation for optimum conditions (Figure 8.10). Figure 8.11 shows the resultant color-coded critical resolution map with one of the optimum simulated chromatograms found at conditions of 50°C and t_G = 60 min. Although the overall resolution was excellent, further method improvements were needed to reduce the long run time. Using the same entered data, a column optimization study was simulated by entering

Table 8.6. Summary of Software and System for HPLC Method Development

System (Vendor) website	Description	Estimated Cost
Software		
DryLab (Rheodyne/LC Resources) www.rheodyne.com	A popular computer modeling software based on a linear solvent strength model in gradient elution. Software incorporated into Waters AMDS. Useful tool for optimizing isocratic and gradient methods and for predicting optimum column configurations.	$5,000 USD for single use
ChromSword Auto (Dr. Galushko) www.chromsword-auto.com	Similar to DryLab and can also work with chemical structures, pK_a, and a preloaded column database. Used as standalone version or together with Agilent or Hitachi HPLC automated method development systems. Able to import data from major chromatography data systems.	$14,000 USD
ACD/Method Development Suite (ACD Labs) www.acdlabs.com	Package of ACD/lab tools for data basing, processing, and optimizing applications for HPLC, GC, and CE and for developing methods based on chemical structures and retained chromatographic knowledge. Package is integrated with ACD/Spec Manager (MS, NMR, IR, UV-Vis) and with ACD database for log P, log D, and pK_a. Able to import data from major chromatography data systems.	
HPLC system for automated method development		
Automated Method Development System, AMDS, (Waters) www.waters.com	Waters Alliance HPLC with PDA and Empower data system with a wizard-based interface software that incorporates DryLab and also imports retention data and uploads methods. Includes peak tracking algorithm and results reporting functions.	$65,000–70,000 USD
Turbo Method Development System (PerkinElmer) www.perkinelmer.com	Series 200 HPLC with PDA and TurboChrom or TotalChrom data system for automated sequential four-solvent grid searches. Mostly for isocratic applications. Includes software for 3-D resolution maps, peak tracking, and robustness evaluation.	$50,000 USD

Column: Waters Symmetry C8
          3.9 x 150 mm, 5 μm
Mobile Ph.: 50% MeOH/water
Flow: 1.0 mL/min at 30°C
Detection: 278 nm

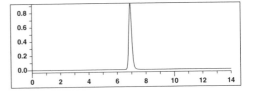

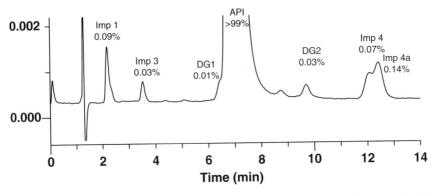

Figure 8.9. HPLC chromatogram of an isocratic composite method for a neutral drug substance.

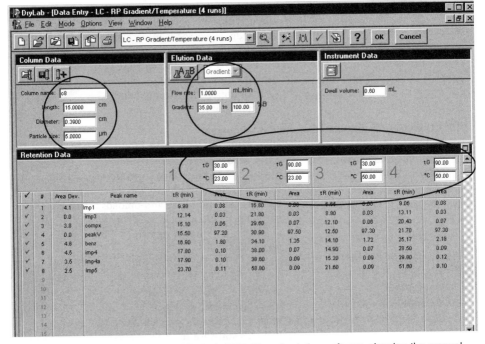

Figure 8.10. Computer screen of DryLab 2000 Plus simulation software showing the manual entry of retention data from results of four gradient runs at two different T and t_G using the 150-mm-long, 5-μm Symmetry C8 column.

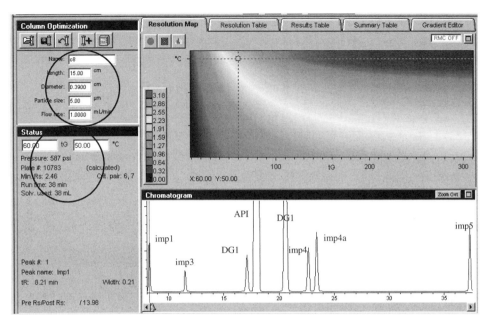

Figure 8.11. A DryLab color-coded resolution map simulated from data from Figure 8.10. The simulated chromatogram at T = 50°C and t_G = 60 min showed excellent resolution but with long analysis time.

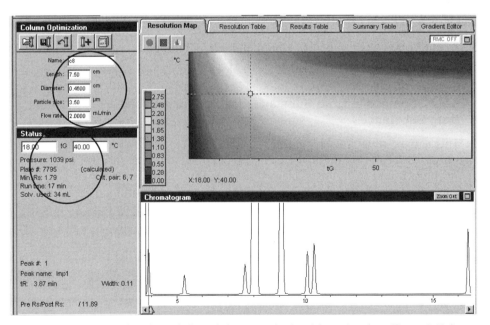

Figure 8.12. A DryLab color-coded resolution map simulated from data from Figure 8.10 for column optimization. The simulated chromatogram at T = 40°C and t_G = 18 min was predicted for a 75-mm-long, 3.5-mm Symmetry C8 column. It showed excellent resolution and run time <22 min.

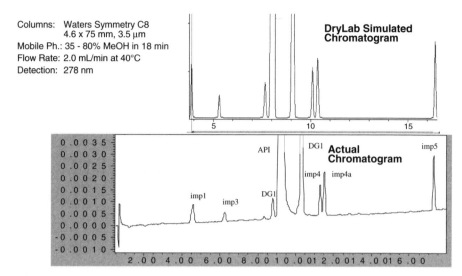

Columns: Waters Symmetry C8
4.6 x 75 mm, 3.5 μm
Mobile Ph.: 35 - 80% MeOH in 18 min
Flow Rate: 2.0 mL/min at 40°C
Detection: 278 nm

DryLab Simulated Chromatogram

Actual Chromatogram

API DG1 imp5 imp4 imp4a imp1 imp3 DG1

Figure 8.13. A comparison of the simulated chromatogram (top) with the actual chromatogram obtained (bottom) by running the predicted column with the predicted optimum condition. Excellent correlations were obtained.

parameters of a shorter Symmetry C8 column packed with smaller particles (75 × 4.6 mm, 3.5 μm) (Figure 8.12). DryLab predicted an optimal separation at 35–80% MeOH in 18 min at 2.0 mL/min and 40°C. The simulated optimal condition has a run time of ~22 min and was confirmed by running an actual confirmatory experiment on an HPLC (Figure 8.13).

8.8.2 Composite Drug Substance Method for a Basic Drug Substance

This case study is a follow-up of the initial development example shown in Figures 8.2–8.4. Conditions used during initial development shown in Figure 8.4 were found to be insufficient due to the co-elution of the API and the immediate synthetic precursor. Varying column temperatures and gradient conditions was ineffective in separating this critical pair. Switching the mobile phase acidifier from 0.1% formic acid to 0.1% trifluoroacetic acid (TFA) was found to yield higher retention and a partial separation. At this point, two polar-embedded columns were evaluated (Waters XTerra RP18 and Supelco Discovery RP-amide), leading to a successful separation using the latter column. The final choice is the Supelco Ascentis RP-amide column, which is a more stable and retentive analog to the Discovery RP-amide column. However, the use of 0.1% TFA gradient was also found to lack sufficient sensitivity due to high gradient shift and baseline noise at 230 nm. By lowering the level of TFA and matching the absorbance of MPA (0.05% TFA in water)

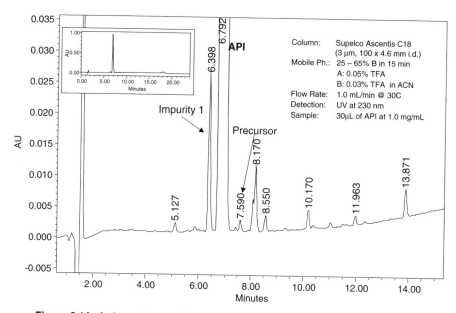

Figure 8.14. A chromatogram of a final method of a basic drug substance sample.

and MPB (0.03% TFA in ACN), a set of optimum conditions with adequate sensitivity (LOQ = 0.05%) were found (Figure 8.14). The validation of this method will be described in Chapter 9, Section 9.5).

8.8.3   Impurity Method for a Drug Product with Two APIs

This case study illustrates the challenging method optimization process for a drug product with two APIs using a "cocktail" test mix consisting of both APIs and all available impurities and potential degradants.[8] Note that the monitoring of process impurities is not required in methods for drug products according to the requirements in the ICH guidelines; however, they are included in method development to assure resolution from the degradants. Both APIs have pK_a at about 8–9 and are highly water-soluble basic salts with low retention at acidic pH. Ion-pairing reagents can improve retention but they tend to reduce sensitivity and are not MS-compatible. Experience with a prior method of a similar product using a high-pH mobile phase[8] led to the development of an optimized condition shown in Figure 8.15. The strategies used were DryLab simulation of T and t_G (Figure 8.16) and the incremental variation of mobile phase pH (Figure 8.17). Mobile phase pH was found to be an extremely powerful but sensitive parameter in controlling resolution of several critical pairs (Figure 8.18). To maintain adequate resolution of all key analytes, the pH of mobile phase A had be maintained between 9.10 and 9.15. Other factors were

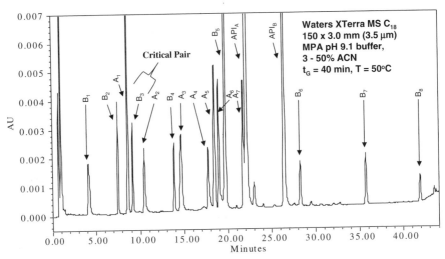

Figure 8.15. A chromatogram of a "cocktail" of two API and some potential impurities for an impurity method of a multi-component drug product. This method uses high-pH mobile phase.

t_G = 20 - 60 min, T = 40 – 55°C, at pH 9.1

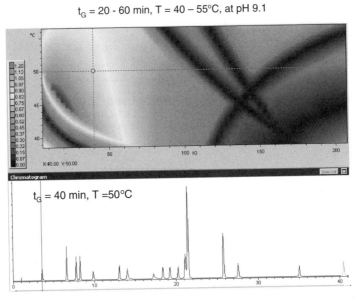

t_G = 40 min, T =50°C

Figure 8.16. A DryLab resolution map from T and t_G. This is a powerful combination for changing selectivity to control band elution in gradient analysis.

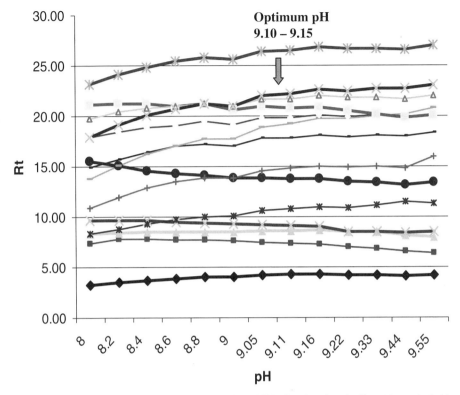

Figure 8.17. A plot of retention data with pH of the MPA showing the significant impact of pH on the separation. Optimum pH was found at pH 9.1.

further evaluated (solvent type and column). Figure 8.19 shows the selectivity effect of substituting acetonitrile by methanol in the mobile phase B. This set of optimum conditions was also used to evaluate five additional modern high-pH C18-bonded phase columns of the same dimensions and particle sizes. Most columns were found to yield similar retention and selectivity. Figure 8.20 shows chromatograms from three of these columns, showing peak tailing problems of two specific impurity peaks. Scrutiny of the structures of these two impurities revealed a highly accessibility to the lone pair of the nitrogen atoms, making them more prone to interaction with residual silanols.

8.9   SUMMARY AND CONCLUSIONS

Method development is a challenging and time-consuming process requiring much experience, creativity, logical thinking, and experimentation. With all the software and automated systems available today, method development is still

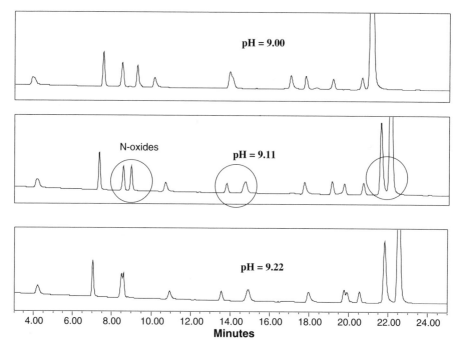

Figure 8.18. Comparative chromatogram at two different pHs near the optimum pH of 9.1. These data predicted some robustness issues toward mobile phase pH since it is close to that of the pK$_a$ of the APIs. A small change of pH would bring dramatic change to the elution order of some analytes. However, it should also be noted that pH is a powerful variable for fine-tuning the resolution of complex samples.

very much a trial-and-error approach, expedited by a logical sequence of generic scouting runs and fine-tuning steps to achieve the requisite resolution and method performance. This chapter describes the generic strategy of first defining method development goals and gathering sample/analyte information, followed by initial method development and method fine-tuning.

The most common initial conditions are reversed phase chromatography using a C8 and C18 column with methanol or acetonitrile and aqueous buffers. A popular approach is the use of a generic broad gradient using acidic mobile phase for overall sample assessment to define the initial separation conditions. It is important to develop MS-compatibile methods to facilitate the identification of unknowns and method troubleshooting. Method fine-tuning typically involves the adjustment of mobile phase factors (%B, buffer, pH, solvent type) or operating parameters (F, T, $\Delta\phi$, t$_G$), which affect selectivity. Modeling software is most useful for mobile phase and column optimization, particularly for gradient and complex separations. A "best practice" in proactive pharmaceutical method development, entitled the "phase-appropriate method development" is described. Several case studies are also included to illustrate the logical sequence and decision process in method development.

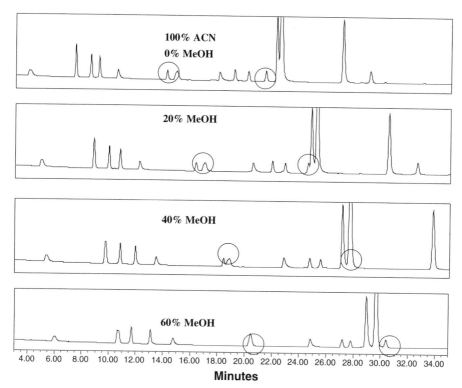

Figure 8.19. Four chromatograms illustrating the effect of organic modifier in MPB, showing the selectivity effect of several impurities.

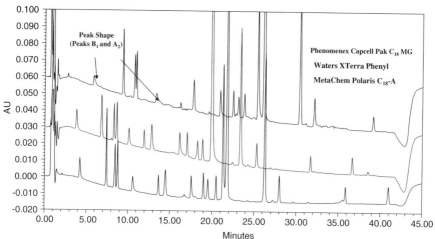

Figure 8.20. Three comparative chromatograms of three different bonded-phase columns using the optimized mobile phase conditions shown in Figure 8.15. The "surprise" finding of peak shape problems with two specific impurities might be attributed to the higher silanophilic activity of the column. Data presented in parts at Eastern Analytical Symposium, Somerset, New Jersey, November 2003.

8.10 REFERENCES

1. L.R. Snyder, J.J. Kirkland, and J.L. Glajch, *Practical HPLC Method Development,* 2nd Edition, Wiley-Interscience, New York, 1997.
2. H. Rasmussen et al., in S. Ahuja and M.W. Dong, eds., *Handbook of Pharmaceutical Analysis by HPLC,* Elsevier, Amsterdam, 2005, Chapter 6.
3. H. Rasmussen, in S. Ahuja and S. Scypinski, eds., *Handbook of Modern Pharmaceutical Analysis,* Academic Press, New York, 2001, Chapter 10.
4. *HPLC Method Development, CLC-70,* (Computer-based Instruction), Academy Savant, Fullerton, CA.
5. "Advanced HPLC Method Development"—a 2-day training course offered by LC Resources, Walnut Creek, CA, www.lcresources.com; M.W. Dong and H. Rasmussen, "HPLC Method Development in Pharmaceutical Analysis,"—a 1-day short course at Eastern Analytical Symposium, November 2005, www.eas.org.
6. S. Mitra, ed., *Sample Preparation Techniques in Analytical Chemistry,* Volume 162. John Wiley & Sons, New York, 2001.
7. C. Choi and M.W. Dong, in S. Ahuja and M.W. Dong, eds., *Handbook of Pharmaceutical Analysis by HPLC,* Elsevier, Amsterdam, 2005, Chapter 5.
8. M.W. Dong, G. Miller, and R. Paul, *J. Chromatogr.* **987**, 283 (2003).
9. U.D. Neue, *HPLC Columns: Theory, Technology, and Practice,* Wiley-VCH, New York, 1997.
10. I. Molnar, *J. Chromatogr. A.,* **965**, 175 (2002).
11. S.V. Galushko and A.A Kamenchuk, *LC.GC Int.* **8**, 581 (1995).
12. Y. Rozenman and J.M. Di Bussolo, *Pharm. Tech.* **10** (Feb 1998).
13. M.W. Dong, R.D. Conlon, and A.F. Poile, *Amer. Lab.* **20(5)**, 50 and **20(6)**, 50 (1988).

9

REGULATORY ASPECTS OF HPLC ANALYSIS: HPLC SYSTEM AND METHOD VALIDATION

Modern HPLC for Practicing Scientists, by Michael W. Dong
Copyright © 2006 John Wiley & Sons, Inc.

9.1 INTRODUCTION

9.1.1 Scope

Because of its high accuracy and precision, HPLC is widely used for the analysis of pharmaceuticals, food products, and environmental samples under regulatory environments to ensure public health and safety. This chapter discusses the various scientific and regulatory issues relating to HPLC analysis, focusing on system qualification, calibration, system suitability, and method validation. Examples from the pharmaceutical industry, including a case study in method validation, are used to illustrate the myriad tools and systems used to ensure HPLC data accuracy. Note that the term *active pharmaceutical ingredients* (API) is synonymous with drug substances.

9.1.2 The Regulatory Environment

The primary regulatory concerns in HPLC analysis relate to data validity, specifically how data can be verified scientifically and documented in a legally defensible manner. While the ultimate goal of regulatory compliance is straight-forward, realization can be complex and time-consuming.[1,2] The manufacture of pharmaceuticals in the United States falls under the regulations of Current Good Manufacturing Practice (cGMP) found in the Code of Federal Regulations (CFR) for finished pharmaceuticals or drug products (21 CFR Part 210 and Part 211). cGMP delineates the requirements for drug manufacturing and production, including those for the personnel, buildings and facilities, equipment, production and process control, packaging and labeling, holding and distribution, records, and reports. cGMP requires a quality documentation system that serves as a communication system to ensure compliance by all personnel involved in the manufacturing, testing, and release of drug product to the public. Documentation takes the form of descriptive documents such as standard operating procedures (SOPs), protocols, and batch records. It also includes data collection documents such as log books and laboratory notebooks; numbering systems such as part numbers, lot numbers, equipment numbers, and method numbers; and data files, including equipment qualification and product files to support accountability and traceability.

To help the reader understand this complex problem, a schematic diagram listing the typical control systems, procedures, tools, and related documentation found in an exemplary cGMP facility (e.g., pharmaceutical production) is provided in Figure 9.1. Compliance with these regulations (shown outside the circle) is enforced by an internal quality assurance (QA) department and by periodic site audits by the U.S. Food and Drug Administration (FDA). All pertinent documents (e.g., SOPs, products and raw materials specifications, analytical methods, and validation reports) require version control, data review, and sign-off approval. Any changes to these controlled documents or deviations from these approved protocols must be handled via "change controls"

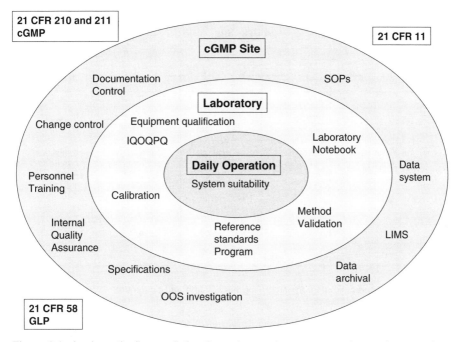

Figure 9.1. A schematic diagram listing the various tools, systems, and procedures used to ensure HPLC data accuracy and regulatory compliance under a cGMP environment. The three circles represent the cGMP organization governing the HPLC laboratory and its daily operation.

or "exception requests" subjected to approval by management and/or the (QA) department. In addition, appropriate personnel qualification and training are considered integral components of cGMP. All drug substances and products must pass published specifications before they can be released for use. All data with regulatory implications are archived to ensure data security and data integrity (to prevent deletion or alteration). All electronic records and signatures are subjected to Title 21 of the Code of Federal Regulations (21 CFR Part 11).

A cGMP analytical laboratory typical operates under a set of SOPs governing the conduct of routine procedures, such as:

- Chemical handling and reagent labeling
- Reference standard qualification and purity assignment
- Data entry into laboratory notebooks
- Writing of controlled documents, i.e., test methods, SOPs, etc.
- Equipment qualification and calibration
- Method validation
- System suitability testing

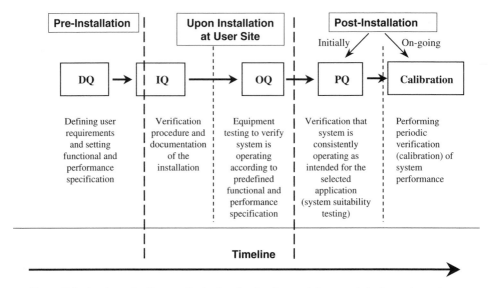

Figure 9.2. A schematic diagram illustrating the timeline and documents in the various stages in HPLC system qualification.

For instance, departmental SOPs define how reference standards are to be certified to confirm identity and purity, how reagents are labeled with expiration dates, and how experimental data are to be entered into laboratory notebooks by the analyst and reviewed by the supervisor. All analytical methods are considered controlled documents and undergo vigorous review and sign-off processes. In addition, a typical three-pronged approach for system and method validation is commonly adopted by the pharmaceutical industry to ensure HPLC data reliability. The first step is *initial system qualification*, followed by *periodic calibration*, which verifies the functional performance of the instrument upon installation and then periodically afterwards. The second step is *method validation*, which verifies the performance of the entire analytical procedure, including sample preparation. The final step is *system suitability testing* (SST), which verifies the holistic functionality of the entire analytical system (including instrument, column, and mobile phase) on a day-to-day basis. These system qualification and method validation approaches are discussed in Sections 9.2–9.4.

9.2 HPLC SYSTEM QUALIFICATION

Equipment qualification is a formal process that provides documented evidence that an instrument is fit for its intended used. The entire process typically consists of four parts: design qualification (DQ), installation qualification (IQ), operational qualification (OQ), and performance qualification (PQ).

After initial qualification, the system is placed in service and is kept in a "qualified state" using a periodic calibration program. In addition, its performance for the specific application is also verified at the time of use using system suitability testing. This section describes the procedures and documentation used for HPLC hardware qualification. Detailed discussions of this subject can be found in books and articles elsewhere.[3,4] The regulations and qualification of computerized data systems and networks with their associated software can be quite elaborate and is described elsewhere.[5-7]

9.2.1 Design Qualification (DQ)

Design qualification (DQ) describes the user requirements and defines the functional and operational specifications of the instrument which are used for OQ testing. Table 9.1 shows an example of a DQ for an HPLC system used for method development.

9.2.2 Installation Qualification (IQ)

Installation qualification (IQ) verifies that the instrument is received in good order as designed and specified by the manufacturer and is properly installed in the user's location. Table 9.2 lists the IQ steps recommended before and during installation. IQ should also include the analysis of a test sample to verify the correct installation of all modules including electrical, fluidics, and data connections.

9.2.3 Operational Qualification (OQ)

Operational qualification (OQ) verifies that the instrument actually functions according to its manufacturer's specifications or, more specifically, within limits defined in the user's DQ. Testing procedures and acceptance criteria in OQ are often similar to those in system calibration shown in Table 9.3. In most laboratories, OQ testing is performed by a service specialist from the manufacturer, while subsequent performance qualification (PQ) and calibration are performed by an in-house metrologist or the user. An OQ protocol is often supplied by the manufacturer for a specific equipment model whereas a calibration protocol is a more generic procedure written for different manufacturers' models, including older equipment with lower performance.

9.2.4 Performance Qualification (PQ)

Performance qualification (PQ) is the process of demonstrating that an instrument can consistently perform an intended application within some predefined acceptance criteria. In practice, PQ testing is synonymous with system suitability testing conducted with specified columns, mobile phases, and test compounds. PQ is performed during initial system qualification or after the system

Table 9.1. An Example of a Design Qualification (DQ) of an HPLC System for Method Development

Design elements	Examples

Intended use
   User requirement
   specification
	Analysis of drug products and substances
	 • Automated analysis of up to 100 samples/day
	 • Limit of quantitation <0.05%
	 • Automated confirmation of peak purity and identity with diode-array detection
	 • Compatible with 2-mm to 4.6-mm i.d. columns

Functional specifications Pump	Quaternary gradient pump with online degassing and a flow range of 0.01–5 mL/min
Detector	UV/Vis diode array, 190–600 nm with 1-nm resolution
Autosampler	>100 sample vials, 0.5–2,000 μL injection volume
Column oven	15–60°C, Peltier
Computer	System control and data acquisition by a computer workstation or a network with remote access
Operational specifications	• Pump precision of retention time <0.5% RSD Composition accuracy <1% absolute • Detector noise, <±2.5 × 10^{-5} AU • Auto sampler precision <0.5% RSD, <0.1 carryover • System dwell volume <1 mL • Instrumental bandwidth <40 μL (4σ)
User instruction	Operating manual
Qualification	Vendor must provide procedures and services for IQ and OQ
Maintenance	Vendor must provide procedures and services for maintenance System must have built-in diagnostic functions
Training	Vendor must provide familiarization and training

is relocated, whereas system suitability testing is conducted on a daily basis or before the instrument is used for regulatory analysis. PQ is often combined into an IQ/OQ protocol during initial installation. System suitability testing is discussed in Section 9.4.

9.2.5 Documentation

Upon completion of equipment qualification, a set of the documentation should be available consisting of the following: DQ document, IQ document, OQ testing procedure, OQ test report, and PQ test procedure and results.

Table 9.2. Steps Recommended for Installation Qualification (IQ)

Before installation
- Obtain manufacturer's recommendations for installation site requirements
- Check site for fulfillment of these requirements (space, electricity, utilities, environmental conditions, and storage space for manuals, software, and log books)

During installation
- Compare equipment as received with purchase order (including software, accessories)
- Check documentation for completeness (manual, IQOQ, certifications)
- Install hardware and software (by service specialists from the manufacturer)
- Power up instrument to perform start-up diagnostic tests
- Run a test sample to verify installation
- Prepare installation report that includes the names and serial numbers of instrument and components, software and firmware versions, actual locations of instrument, manuals, and qualification documentation. Obtain written approval (signatures) from appropriate individuals

9.2.6 System Calibration

"System calibration" refers to the periodic operational qualification of the HPLC, typically every 6 to 12 months in most regulated laboratories. This calibration procedure is usually coordinated with an annual preventative maintenance (PM) program and is performed immediately after PM. A calibration sticker is placed on the instrument to indicate its calibration status and readiness for GMP work. The reader is referred to the principles and strategies behind HPLC calibration criteria published elsewhere.[8] A summary of the calibration procedures and acceptance criteria, including additional procedures recommended for initial operational qualification, is listed in Table 9.3.

9.3 METHOD VALIDATION

Method validation is the process of ensuring that a test procedure is accurate, reproducible, and robust within the specified analyte range for the intended application. Although validation is required by law for all regulatory methods (e.g., cGMP), the actual implementation is somewhat open to interpretation and might differ significantly between organizations. This section describes the highlights of validation parameters (Figure 9.3) and procedures for HPLC methods under ICH guidelines Q2B.[9] The reader is referred to a number of useful books and references for a more detailed discussion of this subject.[10–13] Tables 9.4 and 9.5 list the validation requirements of different method types and at various product development cycles. A case study on the validation of an early-stage composite assay method is used to illustrate the actual process.

Table 9.3. Summary of Typical HPLC Calibration Test Procedures and Acceptance Criteria

HPLC module	Typical test	Procedure (suggested)	Acceptance criteria (suggested)
Detector UV/Vis or PDA	Wavelength accuracy	Measure λ_{max} or maximum absorbance of an anthracene solution (1 µg/mL)	251 ± 3 nm 340 ± 3 nm
Pump	Flow accuracy	Run pump at 0.3 and 1.5 mL/min (65% methanol/water) and collect 5 mL from detector into a volumetric flask. Measure time	<±5%
	Flow precision	Determine retention time RSD of six 10-µL injections of ethylparaben (same as in autosampler test)	RSD <±0.5%
	Compositional accuracy	Test all solvent lines at 2 mL/min with 0.1% acetone/water, step gradients at 0%, 10%, 50%, 90%, and 100%. Measure peak heights of respective step relative to 100% step	±1% absolute
Autosampler	Precision	Determine the peak area RSD of six 10-µL injections of ethylparaben (20 µg/mL)	RSD <±0.5%
	Linearity*	Determine coefficient of linear correlation of injection of 5, 10, 40, and 80 µL of ethylparaben solution	R > 0.999
	Carryover	Determine carryover of peak area from injecting 80 µL of mobile phase following 80-µL injection of ethylparaben	<0.1%
	Sampling accuracy	Determine gravimetrically the average volume of water withdrawn from a tared vial filled with water after six 50-µL injections	50 ± 2 µL
Column oven	Temperature accuracy	Check actual column oven temperature with validated thermal probe	30 ± 2°C 50 ± 2°C
Additional tests recommended during operation qualification			
Detector PDA or UV/Vis	Baseline noise and drift	ASTM method E19.09	<2 × 10^{-3} AU <±2.5 × 10^{-5} AU
Required for systems using Fast LC or <2-mm i.d. columns	Dwell volume	Perform linear gradient with 0.1% acetone/water in 10 min at 1 mL/min without column and measure the intersection of baseline with extrapolate gradient profile (see Fig. 4.5)	<1 mL
	Bandwidth (dispersion)	Measure the 4σ bandwidth of a 1-µL injection of a 0.1% caffeine solution without the column (see Fig. 4.19)	<40 µL

Table adapted and updated from procedures and data found in reference 8 with additional recommendations for OQ testing.
*Linearity of the UV detector and the data system is also verified.

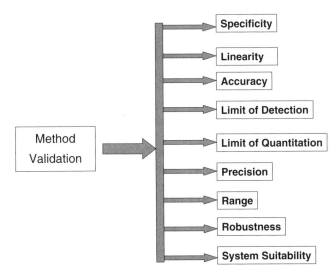

Figure 9.3. A diagram listing of common method validation parameters.

Table 9.4. Validation Requirements for Each Type of Pharmaceutical Analysis Method

Validation parameter	Assay Category I	Assay Category II Quantitative	Assay Category II Limit Test	Assay Category III	Identification
Specificity	Yes	Yes	Yes	Yes	Yes
Linearity	Yes	Yes	No	Yes	No
Accuracy	Yes	Yes	*	Yes	No
Precision	Yes	Yes	No	Yes	No
Range	Yes	Yes	*	Yes	No
LOD	No	Yes	Yes	*	No
LOQ	No	Yes	Yes	*	No
Robustness	Yes	Yes	No	*	No

Table adapted from the USP and ICH guidelines Q2B. *May be required depending on the specific method.

Table 9.5. Typical Validation Requirements at Each Stage of the Product Development Cycle

Product development stage	Validation requirements
Preclinical (prior to human studies)	Specificity, linearity
Initiation of phase 1 clinical studies	Specificity, linearity, range, accuracy, precision, LOQ
Initiation of registration batches, NDA submission	Specificity, linearity, range, accuracy, solution stability, precision, LOQ, robustness
Completion of technology transfer, prior to product launch	Specificity, linearity, range, accuracy, solution stability, precision (repeatability, intermediate precision, and reproducibility), LOD, LOQ, robustness

Note: For Assays, Categories I and II quantitative only.
NDA = New Drug Application.

The reader is also referred to Chapter 6 on parameters and acceptance criteria for the validation of various pharmaceutical analysis methods.

9.3.1 Validation Parameters

Specificity is the ability of a method to discriminate between the intended analyte(s) and other components in the sample. Specificity of the HPLC method is demonstrated by the separation of the analytes from other potential components such as impurities, degradants, or excipients. In addition, stressed samples under forced degradation conditions (acid, base, heat, moisture, light, and oxidation) are used to challenge the method. A method specificity study always includes a demonstration of the noninterference of the placebo sample (Figure 9.4). If a placebo is not available (e.g., drug substance methods), noninterference from contaminants and reagents is demonstrated by running a procedural blank (e.g., injecting the extraction solvent). In addition, specificity to show that there is no co-elution of key analytes is demonstrated by a peak purity assessment using PDA or MS (Figure 9.5), or by comparing the results of the sample to those obtained by a second well-

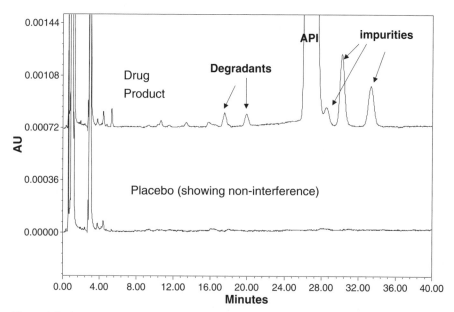

Figure 9.4. An example demonstrating the noninterference of placebo in an impurity testing of a drug product. The upper chromatogram shows the separation of key analytes (API and various impurities and degradants) in an extract of the drug product. The lower chromatogram shows a similar extract of the placebo showing the absence of these key analytes in the placebo extract.

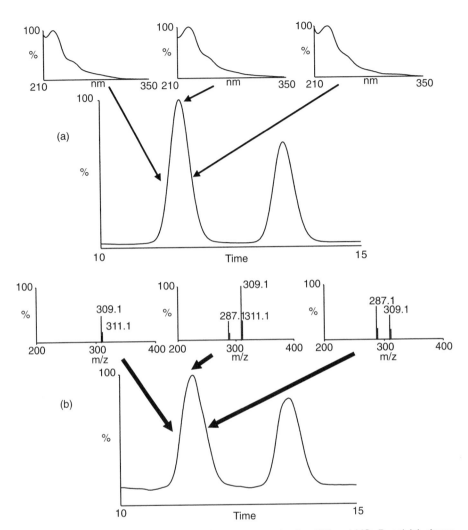

Figure 9.5. Diagram illustrating peak purity assessment using UV and MS. Panel (a) shows that since the UV spectra of the leading, apex, and trailing parts of the first peak look fairly similar visually, one might conclude erroneously that the first peak contains only one single component. Panel (b) shows that the first peak actually contains two components with masses at 311 and 287, respectively. These two components might have similar UV spectra, making UV spectroscopy an insensitive tool for peak purity evaluation. Diagram courtesy of Waters Corporation.

characterized technique. In practice, this secondary technique tends to be another orthogonal RPC method using a different pH or column (Chapter 8).

The **linearity** of a method is its ability to obtain test results that are directly proportional to the sample concentration over a given range. For HPLC methods, the relationship between detector response (peak area or height) and

Table 9.6. Linearity Ranges and Acceptance Criteria for Various Pharmaceutical Methods

Test	Linearity	
	Levels and ranges	Acceptance criteria
Assay and content uniformity	five levels 50–150% of label claim	correlation coefficient, R $R \geq 0.999$ % y-intercept NMT 2.0%
Dissolution	five to eight levels 10–150% of label claim	$R \geq 0.99*$ % y-intercept NMT 5.0%
Related substances	five levels LOQ to acceptance criteria	$R \geq 0.99*$
Cleaning surface validation	five levels LOQ to 20 times LOQ	$R \geq 0.99*$
Bioanalytical	six to eight levels covering the dynamic range	$R \geq 0.99*$

NMT = Not more than. *R can be higher as specified by the company's SOP.

sample concentration (or amount) is used to make this determination. Table 9.6 summarizes the typical linearity levels and ranges as well as the acceptance criteria for various pharmaceutical methods. Figure 9.6 shows the results of a linearity study of an assay method of a drug product at levels of 50, 75, 100, 125, and 150% of label claim. Note that although data show acceptable linear coefficient ($R > 0.999$), the percentage of y intercept is considerably biased (%y intercept ~7%) and does not pass typical assay linearity criterion (e.g., %y intercept of <2%). The likely cause of this bias is UV detector saturation. The use of the typical single-point calibration is not justified if the calibration plot does not pass through zero.

Accuracy is the closeness in agreement of the accepted true value or a reference value to the actual result obtained. Accuracy studies are usually evaluated by determining the recovery of a spiked sample of the analyte into the matrix of the sample (a placebo) or by comparison of the result to a reference standard of known purity. If a placebo is not available, the technique of standard addition is used.

Limit of detection (LOD) is the smallest amount or concentration of analyte that can be detected. There are a number of ways for the calculation of LOD, as discussed in the ICH guidelines on method validation.[9, 10] The simplest method to calculate LOD is to determine the amount (or concentration) of an analyte that yields a peak height with a signal-to-noise ratio (S/N) of 3 (Figure 9.7a).

Limit of quantitation (LOQ) is the lowest level that an analyte can be quantitated with some degree of certainty (e.g., with a precision of ±5%). The simplest method for calculating LOQ is to determine the amount (or concentration) of an analyte that yields a peak with a signal-to-noise ratio of 10

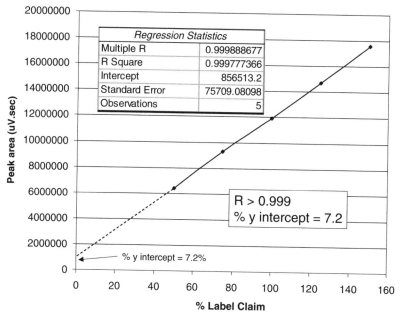

Figure 9.6. Linearity data of an assay method showing the peak area responses at 50, 75, 100, 125, and 150% of the drug substance concentrations. Regression analysis of the data shows a good coefficient of linear correlation ($r > 0.999$) but an obvious bias since the %y intercept is about 7.2%. This bias is typically caused by the nonlinearity of the UV detection in the method.

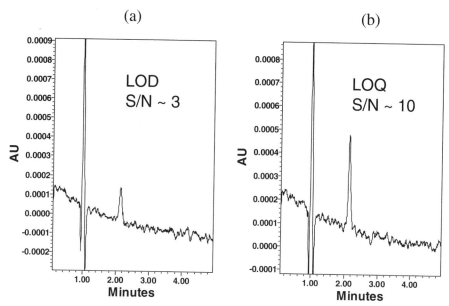

Figure 9.7. Chromatograms showing analyte peaks at limit of detection (LOD, S/N = 3) and limit of quantitation (LOQ, S/N = 10).

Table 9.7. Typical Method Parameters and Range for Robustness Evaluation

Column consistency
  • Three columns packed by bonded phases from three different silica lots
Mobile Phase
  • pH (±0.1–0.2 units)
  • Buffer concentration (±5–10 mM)
  • Percentage organic modifier (±1–2% MPB)
Sample
  • Injection volume or sample concentration
  • Solvent strength for the final solution
Column temperature (±5°C)
Detector wavelength (±3 nm)
Gradient
  • Dwell volume
  • Gradient time (t_G, ±2–5 min)

MPB = mobile phase B or the strong solvent in RPLC. Table compiled from a number of references and by polling several experienced practitioners.

(Figure 9.7b). Thus, LOQ is roughly equal to 3 times of LOD. As noted in Table 9.4 LOQ or LOD determination is required only for high-sensitivity methods involving trace components such as impurity methods or cleaning validation methods.

Method **precision** is a measure of the ability of the method to generate reproducible results. The precision of a method is evaluated for repeatability, intermediate precision, and reproducibility. The reader is referred to Chapter 6 on the precision requirements for various methods.

Repeatability is a measure of the ability of the method to generate similar results for multiple preparations of the same homogeneous sample by one analyst using the same instrument in a short time duration (e.g., on the same day). For instance, method repeatability for pharmaceutical assays may be measured by making six sample determinations at 100% concentration, or by preparing three samples at 80, 100, and 120% concentration levels each.

Intermediate precision, synonymous with the term "ruggedness," is a measure of the variability of method results where samples are tested and compared using different analysts, different equipment, and on different days, etc. This study is a measure of the intra-laboratory variability and is a measure of the precision that can be expected within a laboratory.

Reproducibility is the precision obtained when samples are prepared and compared between different testing sites. Method reproducibility is often assessed during collaborative studies at the time of technology or method transfer (e.g., from a research facility to quality control of a manufacturing plant).

The **range** of an analytical method is the interval between the upper and lower analytical concentration of a sample that has been demonstrated to show acceptable levels of accuracy, precision, and linearity. See examples for the expected validation range for various methods in Chapter 6.

Robustness is a measure of the performance of a method when small, deliberate changes are made to the specified method parameters. The intent of robustness validation is to identify critical parameters for the successful implementation of the method. Robustness is partially evaluated during method development when conditions are optimized to improve resolution and other method performance criteria (e.g., peak shape, sensitivity). Robustness validation is a formalized evaluation of the written method by varying some of the operating parameters within a reasonable range (Table 9.7). These factors can be evaluated one at a time, or preferably by the use of experimental design such as that of Plackett and Burman[14] or other design of experiment (DOE) software packages. Results of the robustness study are used to define system suitability acceptance criteria. Table 9.7 shows the typical parameters and ranges for robustness evaluation.[15]

9.4 SYSTEM SUITABILITY TESTING (SST)

System suitability testing (SST) is used to verify resolution, column efficiency, and repeatability of the analysis system to ensure its adequacy for performing the intended application on a daily basis.[15] According to the latest USP and ICH guidelines, SST must be performed before and throughout all regulated assays. Primary SST parameters are resolution (R), repeatability (RSD of peak response and retention time), column efficiency (N), and tailing factor (T_f). Table 9.8 summarizes guidelines in setting SST limits from the U.S. FDA's Center for Drug Evaluation and Research (CDER) and those proposed by Hsu and Chien,[17] which include recommendations on biologics and trace components. SST acceptance limits should represent the minimum acceptable system performance levels rather than typical or optimal levels. A useful and practical approach is to set SST limits based on the 3-sigma rule gathered from historical performance data of the method.[15]

Table 9.8. Comparison of SST Criteria According to FDA's and Hsu and Chien

SST limits	CDER guidelines	Hsu and Chien's recommendations
Repeatability (RSD) of peak response	≤1.0% for five replicates	≤1.5% general 5–15% for trace <5% for biologics
Resolution (R)	>2.0 in general	>2.0 general >1.5 quantitation
Tailing factor (T_f)	≤2.0	<1.5–2.0
Plate count (N)	>2,000	NA
Capacity factor (k)	>2	2 to 8

NA = not available.

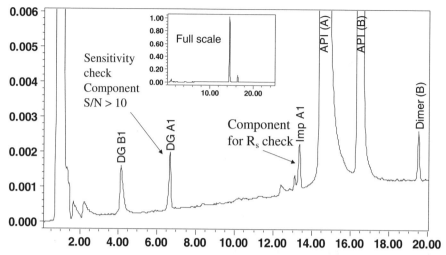

Figure 9.8. HPLC chromatogram of a system suitability solution (SSS) for a challenging impurity testing method for a drug product containing two APIs. This SSS contains both APIs and several key degradants and impurities at their expected concentrations (as retention time markers). One of the component DGA1 present at 0.10% level also serves as a sensitivity check and must meet the acceptance criterion of having S/N > 10.

Recent ICH guidelines indicated that SST must be performed before (initial) and throughout all regulated assays. It is no longer sufficient to assume that the system will function properly during the experiment after passing initial SST. In addition, the use of a single-component calibration solution to check system suitability is not adequate because the system's separation capability is not demonstrated. Rather, the use of system suitability samples (SSS) or resolution test mixtures containing both main components and expected impurities is required. For impurity testing, it is customary to include one of several key impurities in SSS to demonstrate resolution and system sensitivity. An example of an SST chromatogram for an impurity method is shown in Figure 9.8. SSS are analyzed before and interspersed between samples during testing (i.e., five replicate injections of SSS for initial SST and one SSS injection every 10 assay or 12 dissolution samples). Six replicate injections are required if the precision criterion is set above 2% RSD.

If initial SST fails, the analyst should stop the sequence immediately, diagnose the problem, make necessary adjustments or repairs, and re-perform SST. Analysis of actual samples should commence only after passing all SST limits, not only the failed criteria. Most SST failures are traced to problems from the autosampler, pump, column, or mobile phase (see Chapter 10 on troubleshooting hints). If one of the interspersed SSS injections fails, data from all samples after the last passing SSS become invalid and must be repeated.[15]

Table 9.9. Validation Results of a Composite Assay Method (Assay and Impurity) of a Drug Substance

Specificity
  • Resolves API from precursor, impurity 1, and other impurities
  • API peak purity demonstrated by MS and PDA
Linearity
  • API: 5–150% of label claim (eight levels, $R = 1.000$, y% intercept = 1.24%)
  • Precursor: 0.05–2% (five levels, $R = 1.000$)
Accuracy (recovery of spiked, triplicate)
  • API: 98–101% recovered at 70, 100, and 130% levels
  • Precursor: 104–109% recovered at LOQ to 1% levels
Precision RSD ($n = 6$)
  • 0.5% (API at 100%) and 4.5% (precursor at LOQ level)
LOQ = 0.05%, LOD = 0.02%
Solution stability
  • Standard solution stability: >10 days (refrigerated)
  • Mobile phase stable for >14 days
Robustness
  • Demonstrated toward perturbations (T, λ, F, t_G, %TFA) and found to be robust toward SST criteria (precision, T_f, R_s) except for S/N, which is sensitive to λ and %TFA in MPA
  • Resolution of key analytes unaffected by three different lots of column tested

MPA = mobile phase A or the weak solvent in RPLC.

9.5   CASE STUDY ON METHOD VALIDATION

Table 9.9 shows a summary of validation results for the composite test method for a drug substance shown in Figure 9.9. The development process of this particular method is described in Chapter 8, Section 8.8.2. The key analytes of this assay are the API, an impurity eluting (impurity 1) at ~6.4 min that has been identified as an isomer of the API and the immediate synthetic precursor eluting at ~7.6 min.

Method specificity was demonstrated by resolving all impurities from each other and from the API with a focus on two impurities (impurity 1 and precursor). Figure 9.10a illustrates the evaluation of peak purity of the API using the Waters 996 PDA and Empower chromatographic data system. In this peak purity evaluation algorithm, peak purity is indicated by the purity angle being smaller than the purity threshold throughout the peak profile. Another standard technique is to overlay the up-slope and down-slope UV spectra, which show very similar characteristics in Figure 9.10b. Note that unless the impurity peak hidden underneath the main peak have a very different UV spectrum, PDA evaluation is not a particularly sensitivity technique for peak purity assessment. Since the method is MS compatible, LC/MS was also used to evaluation peak purity. The MS spectra of the up-slope, apex, and down-slope of the API peak showed no new m/z ion >1% level in this study. Method linearity, accuracy, and precision studies were performed as listed in Table 9.9

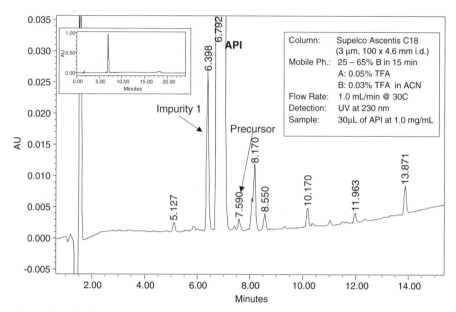

Figure 9.9. HPLC chromatogram of a drug substance assay method showing separation of the API, impurity 1, precursor, and other trace impurities. The inset is the chromatogram at full scale showing a peak height <1.0 AU for the API. Method validation results of this method are shown in Table 9.9.

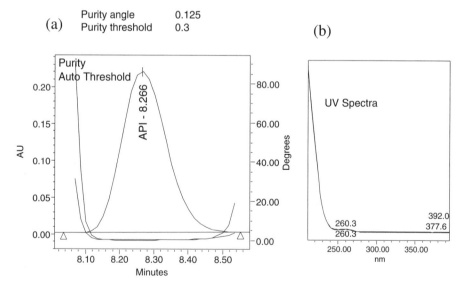

Figure 9.10. (a) Peak purity assessment using the photodiode array detector and Waters Empower purity algorithm by calculating the purity angle and comparing them to the purity threshold. The peak is considered "pure" if the purity angle is lower than that of the purity threshold throughout the peak profile. (b) An overlay of the UV spectra of the up-slope and down-slope portions of the API peak.

Table 9.10. System Suitability Parameters

Parameter	Criterion
Precision of API peak area	RSD ≤1.0%
Precision of API retention time	RSD ≤1.0%
Asymmetry (USP tailing factor) of API	≤2.0
Resolution between API and its isomer	≥2.0
Signal-to-noise of precursor	≥10

for the API and the precursor and were found to be within acceptable limits. A robustness study toward perturbations of five parameters (T, λ, F, t_G, %TFA in MPA) was designed using Design Expert (design of experiment (DOE) software). Eight experimental runs were performed and results were evaluated toward their effects on six system suitability responses (area precision, retention time precision, peak tailing, resolution of the API and impurity 1 (I), and signal-to-noise ratio of the precursor peak (Pre)). Results showed that the method is robust to these variations except for S/N of the precursor peak, which is sensitive to λ and %TFA in MPA. Table 9.10 shows the actual SST acceptance criteria of this early-stage method set from these robustness results.

9.6   COST-EFFECTIVE REGULATORY COMPLIANCE

Regulatory issues in an HPLC laboratory can be quite complex, involving rigorous testing, meticulous documentation, sound scientific judgment, and proper interpretation of regulations to achieve compliance.[1,2] No argument can be made against strict regulatory compliance in production facilities used to manufacture and release drug products for public consumption. However, since the interpretation of the regulations is somewhat inexact and often subjective, the amount of work and documentation required by many organizations has escalated significantly in recent years. Maintaining regulatory compliance has become extremely costly, bureaucratic, and often excessive and inefficient. A case in point is in equipment and software qualification,[4,7] which can often take months to complete and generate hundreds of pages of documentation. While these details may be warranted for the qualification of equipment employing new technologies, most would agree that they might be less meaningful for an off-the-shelf instrument such as a UV spectrometer from a reputable vendor. In response to the need to streamline productivity for faster time to market of new drugs, a delicate balance of laboratory productivity and compliance must be achieved. Many research facilities now adopt a two-tier or risk-based approach, allowing less compliance requirements in laboratories for early-phase development (drug discovery, pre-clinical) and full cGMP or GLP compliance laboratories for late-stage development.

9.7   SUMMARY AND CONCLUSIONS

This chapter discusses regulatory issues in HPLC laboratories with a focus on procedures and requirements for system qualification, calibration, method validation, and system suitability testing. Examples in a cGMP pharmaceutical environment are used to illustrate the various tools and systems used to ensure the degree of HPLC data accuracy necessary to achieve the delicate balance of regulatory compliance and laboratory productivity.

9.8   REFERENCES

1. J.M. Miller and J.B. Crowther, *Analytical Chemistry in a GMP Environment: A Practical Guide*, Jossey-Bass, New York, 2000.
2. L. Huber, *Good Laboratory Practice and Current Good Manufacturing Practice, for Analytical Laboratories,* Agilent Technologies, Walbronn, Germany, 2002.
3. W.M. Reuter, in S. Ahuja and M.W. Dong, eds., *Handbook of Pharmaceutical Analysis by HPLC*, Elsevier, Amsterdam, 2005, Chapter 12.
4. L. Huber, *Validation and Qualification of Analytical Laboratories*, Interpharm, Bufallo Grove, IL, 1998.
5. Title 21 Code of Federal Regulations (21 CFR Part 11), Electronic Records; Electronic Signatures, FDA 2003.
6. W.B. Furman, T.P. Layloff, and R.F. Tetzlaff, *J. AOAC Int.*, **77(5)**, 1314 (1994).
7. L. Huber, *Validation of Computerized Analytical and Networked Systems*, Interpharm Press, Bufallo Grove, IL, 2005.
8. M.W. Dong, R. Paul, and D. Roos, *Today's Chemist at Work*, **10(2)**, 42 (2001). http://pubs.acs.org/subscribe/journals/tcaw/10/i02/html/02dong.html
9. *International Conference on Harmonization (ICH) Q2B*. Validation of Analytical Procedures: Methodology; November 1996. Published in the *Federal Register*, Vol. **62**, No. 96, May 19, 1997, pages 27463–27467.
10. *USP 26/NF 21*; General Chapter <1225>, Validation of Compendial Methods. Rockville, MD. United States Pharmacopoeial Convention, 2003,
11. Guidance for Industry; Analytical Procedures and Method Validation: Chemistry, Manufacturing and Controls Documentation; Draft; August 2000; Center for Drug Evaluation and Research, Center for Biologics Evaluation and Research, FDA, Department of Health and Human Services.
12. M. Swartz and I.S Krull, *Analytical Method Development and Validation,* Marcel Dekker, New York, 1997.
13. A. Lister, in S. Ahuja and M.W. Dong, eds., *Handbook of Pharmaceutical Analysis by HPLC*, Elsevier, Amsterdam, 2005, Chapter 7.
14. R.L. Plackett and J.P. Burman, *Biometrika* **33**, 305 (1943–1946).
15. W.B. Furman, J.G. Dorsey, and L.R. Snyder, *Pharm. Technol.* **22(6)**, 58 (1998).
16. M.W. Dong, R. Paul, and L. Gershanov, *Today's Chemist at Work*, **10(9)**, 38 (2001). http://pubs.acs.org/subscribe/journals/tcaw/10/i09/html/09dong.html
17. H. Hsu and C.S. Chien, *J. Food Drug Anal.* **2(3)**, 161 (1994).

9.9 INTERNET RESOURCES

http://www.fda.gov/cder/guidance/index.htm (U.S. FDA's CDER website containing hundreds of downloadable guidance documents.)

http://www.labcompliance.com (Dr. Huber's website containing useful info on system qualification.)

http://www.ich.org (ICH website)

10

HPLC MAINTENANCE AND TROUBLESHOOTING GUIDE

Modern HPLC for Practicing Scientists, by Michael W. Dong
Copyright © 2006 John Wiley & Sons, Inc.

243

10.1 SCOPE

This chapter gives an overview of HPLC system maintenance and summarizes strategies and guidelines for HPLC troubleshooting. It describes common maintenance procedures that can be performed by the user. Frequently encountered troubleshooting problems are classified into four categories (pressure, baseline, peak, and data). Each problem is described and symptoms and possible solutions are proposed. Several case studies with actual data are used to illustrate the problem diagnosis and resolution process. For a more detailed discussion of the subject, the reader is referred to textbooks,[1-4] magazine columns,[5] training software,[6] manufacturers' operating and service manuals, and myriad web resources (see Section 10.7).

10.2 HPLC SYSTEM MAINTENANCE

Common HPLC maintenance procedures that can be performed by the user are described, including procedures such as replacing check valves, filters, pump seals, detector lamps, flow cells, autosampler sampling syringes, and injector rotor seals. Other more elaborate maintenance tasks such as replacing pump pistons or high-pressure needle seals in some integrated-loop autosamplers are typically handled by the service specialists or internal metrology staff. Most laboratories that work in a regulatory environment have an annual preventive maintenance program for their HPLC systems in which most of the wearable items are replaced. A system calibration procedure typically follows preventive maintenance to verify instrument performance (see Chapter 9).

10.2.1 LC Pump

Common maintenance tasks that can be easily performed by the user are the replacement of solvent filters (sinkers), in-line filters, check valves, and piston seals. Most modern HPLC pumps are designed for easy maintenance, with front panel access to many internal components. Figure 10.1 is a diagram of a slide-out pump unit showing the two pump heads with check valves, the purge valve, the in-line filter, and other components. Procedures for replacing some consumable items are summarized below:

- *Solvent Line Filter (Sinker):* The solvent line sinker, typically a 10-μm sintered stainless-steel filter, can be replaced by direct connection to the Teflon solvent line (Figure 10.2a). The sinkers should be replaced annually. A partially plugged solvent sinker can restrict solvent flow, which can result in poor retention time precision.
- *In-Line Filter:* The in-line filter element (typically 0.5 μm) can be replaced by first disconnecting the connection tubing, opening up the filter body,

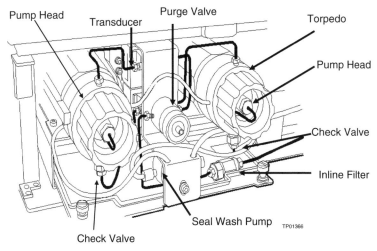

Figure 10.1. Diagram of the inside of the Waters Alliance 2690 pump module after opening the front panel, showing the two pump heads and various components. Note that while the layout on different manufacturers' pumps is not the same, most pumps consist of the same general components and convenience features such as the slide-out tray allowing easy assess for maintenance. Diagram courtesy of Waters Corporation.

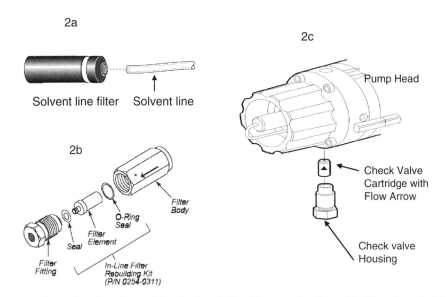

Figure 10.2. (a) The solvent sinker. (b) The in-line filter showing the inside filter element. (c) The pump head with the check valve housing and the check valve cartridge. (a) and (c) Courtesy of Waters Corporation. (b) Courtesy of PerkinElmer, Inc.

and replacing the used filter element with a new unit (Figure 10.2b). A partially plugged in-line filter can cause higher than expected system back pressure.

- *Check Valve:* A check valve can be replaced by first disconnecting the solvent tube, unscrewing the check valve housing, and replacing the inside cartridge with a new unit. The cartridge typically contains a ruby ball and sapphire seat, and must be installed with the flow arrow pointing upward (Figure 10.2c). A used check valve cartridge can often be cleaned by sonication in water, isopropanol, or in 6N nitric acid if heavily contaminated. Higher than expected pressure pulsation is the typical symptom of a bad or contaminated check valve (see Figure 10.3a,b). Note that the inlet check valve is typically more problematic than the outlet check valve.

- *Piston Seal:* The piston seal replacement procedure can be more elaborate and is highly dependent on the particular pump model. A piston seal service kit from the manufacturer should be purchased and the procedure from the service manual should be followed closely for this operation. Typically, the piston should be fully retracted before dismantling the pump head and replacing the seal. A lower than expected system pressure and/or a leak behind the pump head are indications for the need to

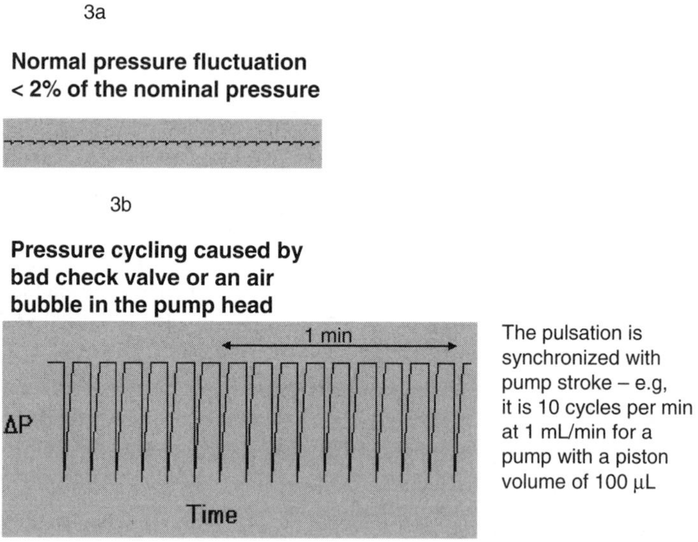

3a

Normal pressure fluctuation < 2% of the nominal pressure

3b

Pressure cycling caused by bad check valve or an air bubble in the pump head

The pulsation is synchronized with pump stroke – e.g, it is 10 cycles per min at 1 mL/min for a pump with a piston volume of 100 µL

Figure 10.3. (a)Pressure profile of a normal HPLC pump showing pressure fluctuation <2%. (b) Pressure profile showing significant fluctuations of a malfunctioning pump due to a bad check valve or a trapped bubble in the pump head. Figure courtesy of Academy Savant.

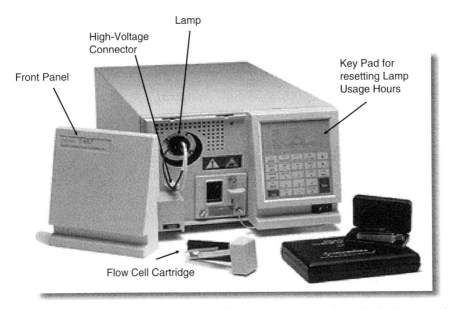

Figure 10.4. A picture of a Waters 2487 UV/Vis absorbance detector with the front panel removed showing the position of the UV lamp (source) and the flow cell cartridge. Picture courtesy of Waters Corp.

replacement of the seal. A lower than expected retention time can also serve as a diagnostic indicator.

10.2.2 UV/Vis Detectors

Modern UV absorbance detectors are designed for easy maintenance and often have front panel access to the lamp and the flow cell. Figure 10.4 shows an example of a modern UV detector with the front panel removed, showing the location of the deuterium lamp and the flow cell cartridge. Both units are self-aligning and do not require any user adjustment upon replacement. Procedures for replacing these items are summarized below:

- *UV Lamp:* Turn off the detector and unplug the power cord. Let the lamp cool for 5 minutes. Disconnect the high-voltage power connector to the old lamp and loosen to remove the securing screws. Replace with the new lamp and reattach the high-voltage connector. Reset the lamp usage hour setting in the detector.
- *Flow Cell:* Disconnect the inlet and outlet solvent tubes, and then loosen to remove the securing screws to the flow cell. Remove the flow cell assembly from the instrument and inspect the flow cell windows for dirt,

particles, contaminants, or window cracks by viewing it against a bright light source. Most flow cells can be disassembled for cleaning or window replacement. Alternately, damaged flow cell may be returned to the manufacturer for repair or reconditioning services.

Some of the optical components (e.g., windows, lens, and mirrors) inside the detector might require cleaning or replacement after several years of use. Indicators for the need to service these optical items are low source energy or low sensitivity performance even after a new lamp has been installed. Occasionally, the monochromator might need adjustment to restore wavelength accuracy. These procedures are best performed by a factory-trained specialist.

10.2.3   Injector and Autosampler

The reliability of HPLC injectors has increased significantly in recent years. The typical wearable item is the injector rotor seal, which should be replaced periodically or when leaks occur. Under normal conditions, a rotor seal can last >10,000 injections. This replacement can be accomplished by most users with some practice by following the instructions. The sampling needle and the sampling syringe are also wearable items that require periodical replacement to restore sampling precision. The sampling syringe is easily accessible and can be replaced by the user. The replacement of the sampling needle can be more elaborate and might require a service specialist on some models. For autosamplers with an integrated-loop design, the high-pressure needle seal also requires annual replacement by a service specialist.

10.3   HPLC TROUBLESHOOTING

"An ounce of prevention is worth a pound of cure," so the best troubleshooting strategy is to prevent problems from occurring by exercising best practices in daily HPLC operation (see Chapter 5) and by performing periodic preventive maintenance. Nevertheless, HPLC problems do occur and often at the most inconvenient times. This section summarizes common HPLC troubleshooting and diagnostic strategies. Discussion focuses on common problems, their associated symptoms, and typical remedial actions. Several troubleshooting case studies are used to illustrate these strategies.

10.3.1   General Problem Diagnostic and Troubleshooting Guide

The following outline is a practical guide for problem diagnosis and troubleshooting.[2,3,5,6]

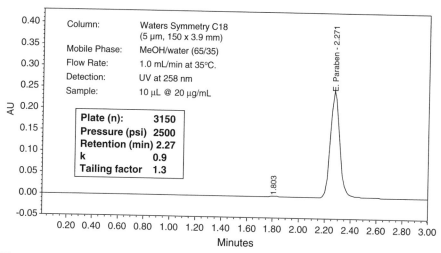

Figure 10.5. HPLC chromatogram of ethyl paraben used as a reference application for problem diagnostics. The inset shows the HPLC conditions and the typical performance parameters such as plate count, pressure, and retention time.

- Verify that a problem exists by repeating the experiment.
- Go back to a documented reference application such as the reference chromatogram with documentation of typically expected retention time, pressure, peak height, baseline noise, and plate count such as the one shown in Figure 10.5. Running this reference application with known good columns and samples and well-prepared mobile phases can often lead to a quick diagnosis. For instance, pump problems are indicated if pressure is lower than expected. A system blockage is indicated if the back pressure is higher than normal. The detector might be faulty if peak heights are lower than expected or if baseline noise is too excessive.
- Isolate problem areas by visual inspection of equipment for leaks and loose cable or tubing connections. Use the built-in diagnostics of each module to locate the problem area (e.g., lamp energy reading for the detector or compression test for the autosampler).
- Other useful troubleshooting strategies include:
  — Performing the obvious or easy things first (i.e., replace check valve if pump pulsation is higher than expected).
  — Troubleshoot one component at a time.
  — Swap the suspect module with a known good one.
  — Consult a service expert or the manufacturer's support line for advice.
  — Do the simple replacement yourself but leave the more elaborate operation to your service specialist.

10.3.2   Common HPLC Problems

Common HPLC problems are caused by component malfunctions (pump, degasser, injector, detector, data system, column), and faulty preparation of the mobile phase or sample preparation. Problems can be categorized into several areas:

- Pressure problems
- Baseline problems (chromatogram)
- Peak problems (chromatogram)
- Data performance problems

Each area is discussed with typical symptoms and their resolution.

10.3.2.1 Pressure Problems and Causes   *Pressure too high:* Higher than expected system pressure is caused by partial blockages in system components such as filter, guard column, column, connection tubing, injector, or detector tubing. The solution is to isolate the source for the partial blockage by disconnecting one component at a time. For instance, if higher than expected pressure is experienced after disconnecting the column, the blockage is occurring upstream in the injector or the pump. If pressure is still high after disconnecting tubing to the injector, the problem then lies in the pump. The most likely location of blockage is the column that is packed with small particles. High column back pressure can be remedied by back-flushing or replacing the inlet frit or the column. A high-pressure problem should be investigated early, before total blockage occurs.

Pressure too low: Lower than expected system pressure is caused by leaks (piston seal, column connections, injector), pump malfunctions (lost prime, air bubbles in pump head, vapor lock, faulty check valves, broken piston), or inadequate solvent supply (empty solvent reservoir, plugged solvent sinker, bent solvent lines, or wrong solvent mixture). Problem diagnostics can be made by visual inspection for leaks and by monitoring the pressure reading of the pump.

Pressure cycling: Pressure cycling is caused by air bubbles in check valves (remedied by degassing the solvent), malfunctioning of check valves or pulse dampeners, or partial system blockages. A typical pressure profile of a modern pump is shown in Figure 10.3 with a pressure fluctuations specification of <±2% of the nominal pressure. A malfunctioning check valve is typically the result of a trapped air bubble or contamination of the ball and seat mechanism causing incomplete closures. It usually manifests itself by significantly increased pressure fluctuations that are synchronous to the pump strokes (Figure 10.3). An air bubble can be dislodged by degassing all solvents and by purging the pump with methanol or isopropanol that wets the check valve more effectively. Malfunctioning check valves should be dismantled and cleaned, or replaced by a new unit.

10.3.2.2 Baseline Problems (Chromatogram) Common baseline problems are illustrated with several diagnostic examples shown in Figures 10.6 and 10.7. The description of each symptom and its possible remedial action is as follows.

Noisy baseline: Short-term detector noise can be estimated by using an expanded scale and measuring the peak-to-peak signal fluctuation of the baseline. The noise of a modern UV detector should be close to the published specification or $\pm 1 \times 10^{-5}$ AU. Noisy baseline such as the one shown in Figure 10.7a is typically caused by low energy of an aging UV lamp which should be replaced. Detector noise can be caused by a large air bubble trapped in the flow cell or pressure fluctuations caused by a small leak in the flow cell.

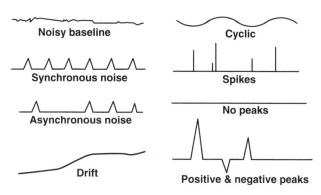

Figure 10.6. Examples of various baseline problems. Figure courtesy of Waters Corporation.

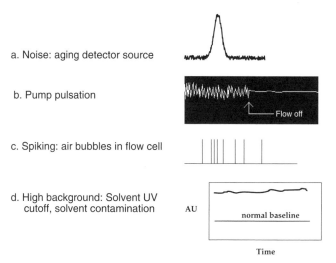

Figure 10.7. Examples of various baseline problems and possible causes.

A noisy baseline of an RI detector can be caused by inadequate mobile phase degassing or temperature thermostatting. However, low light energy can also be caused by a contaminated detector flow cell or high UV absorbance of the mobile phase (Figure 10.7d).

Synchronous and asynchronous noise: Cyclical baseline noise in sync with pump strokes is typically pump-related (Figure 10.3b). For instance, a flow rate of 1 mL/min delivered by a pump with a piston volume of 100 µL, will operate at 10 cycles per minute. One definitive test for pump-related noise is to check if the noise is eliminated after the pump is turned off (Figure 10.7b). This synchronous noise can be caused by pump malfunctions (small air bubble in pump head, check valve problems, broken plunger, or faulty pulse dampener). Inadequate mixing of the mobile phase is also a possible cause if solvent blending at the pump is used. This can be remedied by using a larger mixer or switching to a premixed mobile phase. Synchronous detector noise (which is not in sync with the pump and occurs at lower frequency) can be caused by electrical sources such as ground loop problems, power/voltage fluctuations, or electrical shielding problems. These situations might be difficult to diagnose and remedy. Asynchronous noise is likely caused by the detector, such as leaks, loose detector connections, or electric problems.

Baseline drift: Baseline drift in an UV absorbance detector often reflects change of the energy output of the UV lamp with time. Baseline drift is reduced significantly in dual-beam absorbance detectors because the inherent fluctuation of the lamp energy is compensated. A specification of $\sim 1 \times 10^{-4}$ AU/ hour is typical. Baseline shifts associated with gradient analysis are quite normal. They are caused by the difference of absorbance and/or refractive index of the initial and final mobile phase (see case study in Figure 10.15). These gradient shifts can be minimized by balancing the absorbance of the two mobile phases (MPA and MPB) and/or by using a flow cell designed to minimize the effects of refractive index changes (e.g., tapered flow cell). Nonspecific drifts can be caused by strongly retained peaks slowly bleeding off a contaminated column. These problems can be remedied by purging the column with a strong solvent until the baseline is stable. Drift can also be caused by temperature fluctuations or small leaks in the flow cell. Drifts caused by thermal effects are particularly prominent in refractive index, electrochemical, and conductivity detectors. Cyclic baseline drift can be caused by cyclical thermal effect such as periodic cooling from a ventilation source or cyclical voltage fluctuations such as those caused by the periodic current draws from a high-wattage compressor on the same circuit as the HPLC detector.

Spikes on baseline: Spikes on baseline (Figure 10.7c) are caused by air bubbles out-gassing in the detector flow cell. They can be eliminated by mobile phase degassing and/or by placing a pressure restrictor in the detector outlet (e.g., with 50-psi back pressure). Spikes can also be caused by poor signal wire connections (loose or damaged wiring) or malfunction of the detector or the data system.

10.3.2.3 *Peak Problems (Chromatogram)*   Peak problems such as poor peak resolution, broad peaks, split peaks, tailing or fronting peaks, and/or extra peaks are most often caused by the column and its interaction with the mobile phase and the sample. "No peaks in the chromatogram" could be due to a number of causes including the injector not making injection, a pump not delivering flow, big leaks, a dead detector, a mis-wired data system, wrong mobile phase, a particularly retentive or adsorptive column, or a bad/wrong sample. The best procedure to diagnose such problems is to go back to a reference test conditions such as those show in Figure 10.5. If the reference conditions can be duplicated successfully, thereby inferring that the system is functioning properly, then the troubleshooting effort can be directed to the specific application conditions (column, mobile phase, sample, etc.).

Broad peaks and split peaks: Abnormally broad peaks and split peaks are indications of degraded column performance caused by sample contamination, partially blocked inlet frit, or column voiding (Figures 10.8 and 10.9). They are remedied by back flushing the column, refilling the column inlet with packing, or, more often, by replacing the column. Note that broader than expected peaks for early eluting peaks are often caused by extra-column bandbroadening from the HPLC system (Figure 10.10). These extra-column bandbroadening effects are particularly deleterious for small-diameter or short Fast LC columns. Anomalous peak shapes can also be caused by injecting samples dissolved in solvents stronger than the mobile phase (Figure 10.11). If possible, the solvent for dissolving the sample should be of a weaker strength than the mobile phase. If strong solvents must be used, the volume should be kept

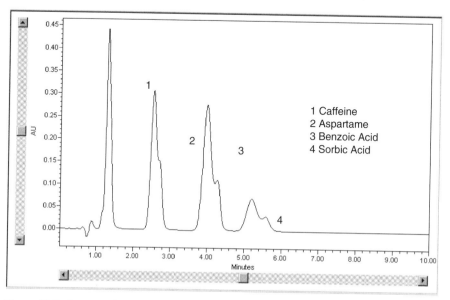

Figure 10.8. An example of chromatographic peak splitting possibly caused by column voiding.

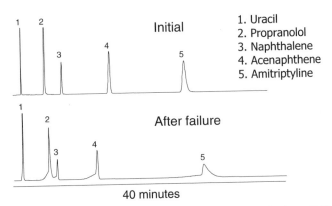

1. Uracil
2. Propranolol
3. Naphthalene
4. Acenaphthene
5. Amitriptyline

Figure 10.9. Comparative chromatograms of a good and a failed silica-base C18 column due to exposure to high pH mobile phase. Severe peak fronting indicated channeling of the packed bed while peak tailing of the basic analyte (amitriptyline) indicated increase of silanophilic activity due to loss of endcapping or bonded groups. Figure courtesy of Waters Corporation.

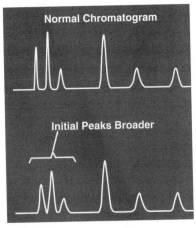

Figure 10.10. An example of peak broadening caused by instrument dispersion (extra-column band-broadening), which affects the early eluting peaks more severely. Figure courtesy of Academy Savant.

small (e.g., <5 µL) to prevent peak broadening or peak splitting of the early eluting peaks. Peak splitting can also result when the sample pH is different from the pH of the mobile phase.

Fronting and tailing peaks: Fronting peaks are caused by column overload (sample amount exceeding the sample capacity of the column), resulting in some of the analytes eluting ahead of the main analyte band. Tailing peaks are caused by secondary interaction of the analyte band with the stationary phase. A common example is the tailing peaks of basic analytes caused by strong

Peak Shape vs Sample Solvent Strength

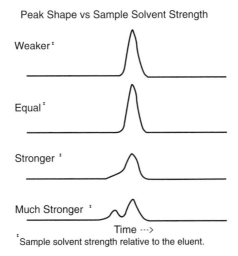

Weaker[x]

Equal[x]

Stronger[x]

Much Stronger[x]

Time --->

[x] Sample solvent strength relative to the eluent.

Figure 10.11. An example of peak broadening and splitting caused by injecting sample dissolved in solvent stronger than that of the initial mobile phase. Splitting occurs since some analyte molecules are eluted prematurely by the stronger sample solvent. Figure courtesy of Academy Savant.

interaction with the acidic residual silanol groups of the silica-based bonded phases. Peak tailing of a basic analyte can be reduced by adding an amine modifier (0.1% trimethyl amine) to the mobile phase or by replacing the column with lower silanophilic activity (see Chapter 3). Asymmetric peak shapes can also be caused by chemical reactions or isomerization of the analyte during chromatography.

Negative and positive peaks: Negative peaks or dips in the chromatogram are common near the solvent front. They are caused by refractive index changes or by sample solvents with less absorbance than the mobile phase. These baseline perturbations are usually ignored by starting data integration after the solvent front. If all peaks are negative, it might be an indication of the wrong polarity in the detector wire connection unless an indirect UV detection technique is used. Negative peaks are most commonly encountered when using an RI detector.

Ghost peaks and extra peaks: Ghost peaks associated with a blank gradient (running a gradient by injecting just the sample solvent) are usually caused by trace contaminants in the weaker mobile phase (starting mobile phase A) and can be minimized by using purified reagents and by exercising stringent mobile phase preparation precautions (See Figure 10.12 and Chapter 5 in trace analysis). Unexpected extra peaks, particularly those which are unusually broad, are typically caused by late eluting peaks from prior injections (Figure 10.13). Some of these peaks can be so broad that they are often mistaken for baseline drifts.

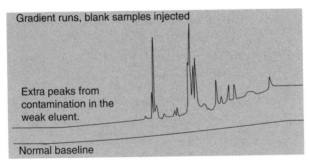

Figure 10.12. An example of many ghost peaks in a gradient analysis of a blank sample. These peaks are impurities or contaminants in the initial weaker mobile phase (mobile phase A), which are concentrated and eluted during gradient analysis. A more normal baseline can usually be established by preparing a fresh batch of MPA with purified reagents. Figure courtesy of Academy Savant.

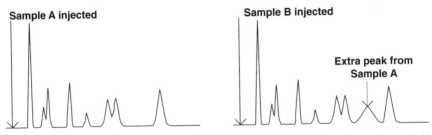

Figure 10.13. An example of an unexpected broad peak caused by the elution of a highly retained peak from the previous injection. Figure courtesy of Academy Savant.

10.3.2.4 Data Performance Problems   Data performance problems such as poor precision and inaccurate results might be difficult to diagnose. Data problems can stem from system malfunctions or problems relating to the mobile phase, method calibration, or sample preparation. Poor system precision of retention time and peak area, and their possible causes and remedies, are covered in Chapter 5, Section 5.5.6 and in several published articles.[7,8] Poor method accuracy can be caused by peak co-elution (peak purity), sample stability, sample preparation/recovery problems, matrix interferences, equipment calibration, or integration problems (Figure 10.14). One excellent guideline to follow is to always print reports with chromatograms that show the peak baselines to check area integration. Data problem situations typically require the insights of an experienced practitioner for correct diagnosis and remedy. They should be investigated further by eliminating each cause as illustrated in some of the case studies below.

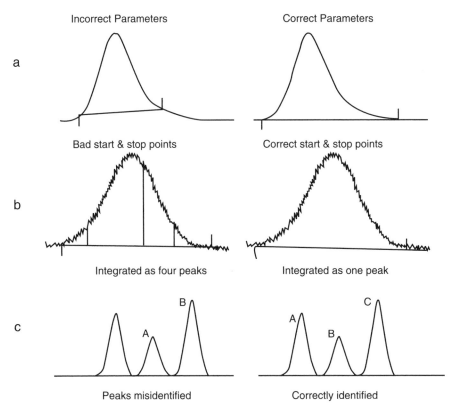

Figure 10.14. Example of various peak integration problems caused by setting incorrect integration parameters such as bad baseline and area thresholds (a and b) or wrong retention time windows (c). Figure courtesy of Waters Corporation.

10.4 CASE STUDIES

Several case studies concerning baseline and data performance problems, as well as troubleshooting examples, are selected here to illustrate problem diagnosis and resolution.

10.4.1 Case Study 1: Reducing Baseline Shift and Noise for Gradient Analysis

Figure 10.15 illustrates a gradient baseline shift problem encountered during the method development for a method to analyze impurities of a drug substance using mobile phase containing 0.1% trifluoroacetic acid (TFA) (case study in Section 8.8.2). Due to the absorbance of TFA in the far UV region (200–230 nm) (Figure 10.15c), considerable baseline shift (0.1 AU at 230 nm)

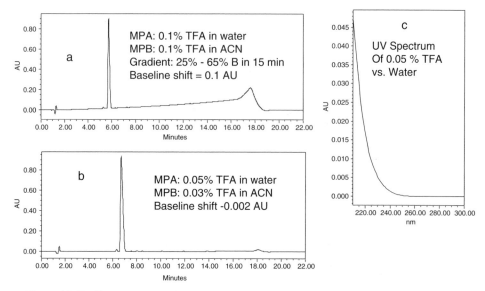

Figure 10.15. Chromatograms illustrating gradient shift problems at 230 nm encountered during method development of an impurity test method for a drug substance (Section 8.8.2). The UV spectrum of 0.05% TFA vs. water is shown in the inset, showing considerable absorbance at 230 nm.

(Figure 10.15a) and substantial baseline fluctuations (Figure 10.16a) were observed. These baseline problems were substantially reduced by lowering the concentration of TFA in both the starting mobile phase (MPA) and final mobile phase B (MPB) from 0.1% to 0.05% (Figures 10.15b and 10.16b). Further reduction of the gradient shift and baseline fluctuation can be achieved by adjusting the concentration of TFA in MPB until it matches the absorbance of MPA of 0.05 TFA in water (i.e., by balancing the absorbance of MPB with MPA). Note that MPB used in Figure 10.15b contains only 0.03% TFA in acetonitrile due possibly to the spectral wavelength shift of TFA in acetonitrile.

10.4.2   Case Study 2: Poor Peak Area Precision Encountered During HPLC System Calibration

Two real-world examples are used to illustrate how to resolve this common problem encountered during the calibration of an autosampler (Chapter 9). In Table 10.1, the precision of peak area was found to be >1% RSD, which was above the acceptance criterion of 0.5% RSD. Repeating the experiment showed similar precision results. Since a worn sampling syringe of the autosampler is the most common cause for poor precision, it was replaced. The peak area precision was found to be ~0.3% RSD after this replacement.

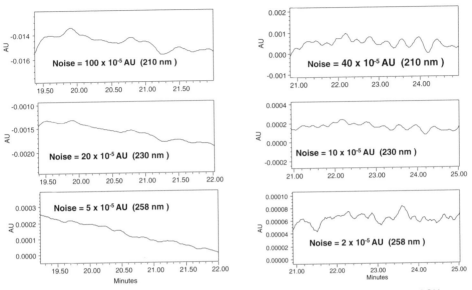

Figure 10.16. Baseline fluctuations from blending 0.1% TFA in water and 0.1% TFA in ACN vs. that of 0.05% TFA in water and 0.03% TFA in ACN at different wavelengths (210, 230, and 258 nm). The magnitude of baseline noise is dependent on monitoring wavelength and is caused by pump blending and the absorbance of TFA.

Table 10.1. Peak Area Precision Data Before and After Replacing the Sampling Syringe of the Autosampler

Run	Peak Area	Peak Area After replacing syringe
1	1,394,782	1,419,308
2	1,432,841	1,421,143
3	1,445,966	1,409,706
4	1,422,240	1,410,025
5	1,422,555	1,410,880
6	1,410,891	1,408,446
7	1,418,816	1,409,981
8	1,404,683	1,412,092
9	1,425,581	1,410,867
10	1,428,552	1,414,600
Average	1,420,691	1,412,705
RSD%	**1.02%**	**0.31%**

Table 10.2. Peak Area and Retention Time Precision Data Before and After Replacing the HPLC Column

	With aged column			With new column	
Run	t_R (min)	Peak area	Run	t_R (min)	Peak area
1	2.350	998,451	10	2.454	1,159,680
2	2.371	1,014,657	11	2.458	1,181,497
3	2.386	1,023,041	12	2.465	1,200,950
4	2.401	1,036,924	13	2.464	1,219,282
5	2.414	1,055,259	14	2.465	1,233,082
6	2.427	1,073,619	15	2.470	1,248,339
7	2.437	1,097,006	16	2.470	1,260,375
8	2.444	1,116,404	17	2.471	1,273,832
9	2.448	1,139,731	18	2.476	1,284,628
Avg.	2.437	1,145,375	Avg.	2.483	1,365,489
RSD%	1.56%	**8.45%**	RSD%	0.20%	**0.12%**

In the second scenario shown by the data in Table 10.2, poor precision was encountered in both the retention time and peak area of an analyte. Scrutiny of the data showed that both retention time and peak area tended to trend upward. Troubleshooting the autosampler by replacing the sampling syringe yielded no improvements. The poor precision of retention time hinted to possible column problems. The column was therefore replaced and acceptable precision was restored.

10.4.3 Case Study 3: Poor Assay Accuracy Data, an Out-of-Specification Investigation

Table 10.3 shows a set of data generated during the assay of five pharmaceutical suspension samples (in duplicate preparation) prepared for clinical testing. Data showed at least two suspected out-of-specification (OOS) results because they were beyond the specification range of 90–110% of label claim. Data review indicated that the HPLC system was functioning properly and passed system suitability testing. Subsequent laboratory investigation indicated that incomplete extraction of drug substance from the suspension was likely due to the relatively low solubility of the drug in the extraction solvent. In a subsequent study, the two retained suspect OOS sample extracts were sonicated and analyzed and found to have assay values of over 98%. The extraction procedure of the analytical method was thus revised to include an additional vortexing step prior to a longer ultrasonication extraction. After this method revision, no similar OOS results were obtained.

Table 10.3. Assay Data of Five Separate Samples (in Duplicate Preparation) of a Drug Suspension Containing a Low Solubility Drug Substance

Sample	Preparation	% Label claim
1	a	99.5
	b	100.2
2	a	99.1
	b	99.6
3	a	100.8
	b	75.7**
4	a	100.0
	b	99.5
5	a	84.3**
	b	91.7*

**Suspect out-of-specification and *out-of-trend samples.

10.4.4 Case Study 4: Equipment Malfunctioning

Case study 4 shows several examples of problems caused by equipment malfunctions and their subsequent diagnosis and solution. The first one involved a situation of poor retention time reproducibility of a gradient assay. It involved the analysis of a complex natural product, using a narrowbore column (2-mm i.d.) at 0.5 mL/min. System suitability test showed retention times to be erratic and could vary by 1–2 minutes without any obvious trends. Flow rate accuracy was found to be acceptable, however, the compositional accuracy test failed (see Chapter 9 on HPLC calibration). The tentative diagnosis was that of a malfunctioning of the proportioning valve. After its replacement, the retention time precision performance was re-established.

The second example of instrumental malfunctions was encountered during an isolation of an impurity of a drug substance using a preparative HPLC system. The symptom was a classic example of detector baseline "spiking" (Figure 10.7c) caused by bubble formation in the flow cell. Pump blending was used since gradient elution was needed for this isolation. Since the system was only equipped with helium degassing, helium was turned on to degas the solvent and a back-pressure restrictor (50 psi) was placed at the detector outlet. Surprisingly, detector spiking actually became worse. It was later found that a tank of argon was used for degassing by a new analyst who replaced an empty helium tank. Though argon, like helium, is also an inert gas, it is not effective for solvent degassing due to its relatively high solubility in the mobile phase. The spiking ceased after actually degassing the mobile phase with helium.

The last example showed a case of a mysterious baseline rise occurring at exactly 30 min (encountered repeatedly like clockwork) during a long

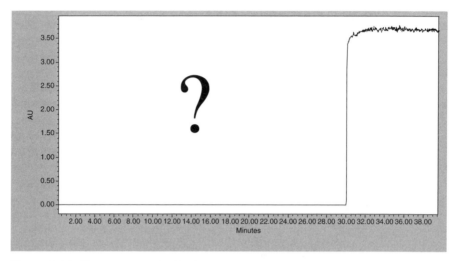

Figure 10.17. Chromatogram of a blank injection showing the case of a mysterious baseline rise at exactly 30 minutes. Note that the baseline after 30 minutes was at 3.5 AU and very noisy.

Table 10.4. Summary of Symptoms of Common HPLC Problems and Probable Causes

	Symptom	Most probable cause
Pressure	Too high	Column, filter
	Too low	Leaks from column, pump
Baseline	Noisy	Old lamp, mobile phase
	Drifting (wandering)	Degassing, mobile phase, contaminated column
Peak	No peak	Detector, injector
	Bad peaks	Column, mobile phase, bad column connections
Data performance	Retention time	Degassing, column, temperature changes
	Peak area	Autosampler, column, integration, lamp
	Accuracy	Sample prep, autosampler

gradient assay. This baseline rise was occurring even with a blank injection and problems with the samples were therefore unlikely (Figure 10.17). Investigation revealed a programming error of the system control method in which the detector lamp was turned off at 30 min. The analyst had copied the method from a different assay method which included a lamp shutdown at 30 min after injection of the last sample. Deleting that lamp shutdown step eliminated the problem.

10.5   SUMMARY AND CONCLUSION

Table 10.4 summarizes the symptoms of common HPLC problems and the most probable causes. Developing HPLC troubleshooting skills often takes many years of operating experience and a working understanding of the principles of the instrument as well as considerable patience to eliminate all the typical causative factors.

10.6   REFERENCES

1. L.R. Snyder and J.J. Kirkland, *Introduction to Modern Liquid Chromatography,* John Wiley & Sons, New York, 1979.
2. L.R. Snyder, J.J. Kirkland, and J.L. Glajch, *Practical HPLC Method Development,* 2nd Edition, Wiley-Interscience, New York, 1997.
3. J. Dolan and L.R. Snyder, *Troubleshooting LC Systems*, Humana Press, Totowa, NJ, 1989.
4. P.C. Sadek, *Troubleshooting HPLC Systems: A Bench Manual*, John Wiley & Sons, New York, 2000.
5. J. Dolan, "*HPLC Troubleshooting*" columns in *LC.GC Magazine*.
6. *Troubleshooting in HPLC, CLC-70,* (Computer-based Instruction), Academy Savant, Fullerton, CA.
7. M.W. Dong, *Today's Chemist at Work,* **9(8)**, 28 (2000).
8. E. Grushka and I. Zamir, in P. Brown and R. Hartwick, eds., *Chemical Analysis,* Volume 98, Wiley Interscience, New York, 1989, p. 529.

10.7   INTERNET RESOURCES

https://www.mn-net.com/web/MN-WEB-mnweb.nsf/WebE/CWIK-4P5F9K
http://www.metachem.com/tech/Troubleshoot/troubleshoot.htm
http://kerouac.pharm.uky.edu/asrg/hplc/troubleshooting.html
http://www.waters.com/WatersDivision/pdfs/WA20769.pdf
http://www.chromatography.co.uk/Techniqs/hplc/trouble1.html
http://www.sigmaaldrich.com/Graphics/Supelco/objects/4500/4497.pdf
http://www.dq.fct.unl.pt/QOF/hplcts1.html
http://www.forumsci.co.il/HPLC/topics.html#Trouble
http://www.rheodyne.com/support/product/troubleshooting/index.asp

11

MODERN TRENDS IN HPLC

HPLC is a mature analytical technology that has undergone continuous refinements during the last four decades. This chapter summarizes the present status and future trends in three broad areas of HPLC technology—columns, instru-

Modern HPLC for Practicing Scientists, by Michael W. Dong
Copyright © 2006 John Wiley & Sons, Inc.

mentation, and methodology. The format of this chapter was inspired by John Naisbitt's 1988 best-selling book, *Megatrends*,[1] which described the 10 new directions transforming our lives at that time. It seems fitting to use this same term in the final summary chapter to describe the 10 most important trends in liquid chromatography that impact our lives as practicing chromatographers today. For further information, the reader is referred to Chapters 3, 4, 8, and 9 of this book and of their associated references.

11.1 COLUMNS: SHORTER AND NARROWER PACKED WITH SMALL PARTICLES

HPLC columns are becoming ever narrower and shorter. The length of the "standard" column has decreased during the past several decades from 250–300 mm to 100–150 mm. This was accompanied by a concomitant decrease in particle size of the packing materials from 10 μm to 3–5 μm. The use of sub-2-μm packing is emerging. For routine analysis, the "standard" 4.6-mm i.d. column remains popular, with an increasing usage of narrower columns in the 2–3 mm range. For LC/MS analysis, the use of 2-mm i.d. columns predominates. For sample limited applications (e.g., biomedical research, proteomics), fused silica capillary columns (micro and nano LC columns) are typical. This trend of using shorter and narrower columns is expected to accelerate as low-dispersion instruments become more available for routine analysis.

11.2 COLUMN PACKING: NOVEL BONDED PHASES

HPLC columns packed with high-purity, silica-based bonded phases continue to dominate the market. Modern columns yield more symmetric peaks for basic analytes (less silanophilic activity) and have better batch-to-batch reproducibility and longer lifetimes. Improved bonding chemistries have widened the usable pH range from 2–8 to 1.5–10 or more. Although C8- and C18-bonded phases remain the most common, other phases have become quite popular, including phenyl, cyano, and several polar-embedded phases (e.g., amide, carbamate).

11.3 PUMPS

Modern pumps have higher reliability (longer seal life) and precision at lower flow rates (<0.2 mL/min) brought forth by innovations such as the dual pistons in series design, variable stroke mechanism, micro pistons, and better check valves. Low-pressure mixing quaternary pumps have become standard equipment in research laboratories for method development, while high-pressure mixing pumps with its lower dwell volumes are popular for LC/MS, high-throughput screening, and micro LC applications.

11.4 AUTOSAMPLERS

Autosamplers have made significant progress in reliability and performance (sampling precision and carryover). Two types of autosamplers remain dominant: the X-Y-Z robotics and the "integrated-loop" designs. Precision levels of <0.2% relative standard deviation (RSD) and low carryover of <0.05% are routine for modern autosamplers. Many autosamplers are capable of fast operation (up to 3–6 injections/min) and sampling from 96-well microplates, important for high-throughput screening (HTS) and bioanalytical applications. Most have optional Peltier cooling sample trays.

11.5 DETECTORS

The dual-beam variable wavelength UV/Vis absorbance detector remains the primary detector for routine analysis. Sensitivity and linearity performance have improved significantly in recent years. Noise specifications of $\pm0.25 \times 10^{-5}$ AU and linearity of upper limit exceeding 2 AU are achievable with newer models. Typical spectral bandwidth ranges from 5 to 8 nm. The lifetime of the deuterium lamp has improved to 1,000–2,000 hours. Most detectors have features such as self-aligned sources and flow cells, leak sensors, and built-in holmium oxide filters for easy wavelength accuracy verification. Many have dual- or multiple-wavelength and ratio plot output capabilities.

For method development and applications requiring peak identification, the photodiode array (PDA) detector is indispensable. Most PDAs are single-beam instruments using single photodiode arrays with either 512 or 1,024 pixels and capable of spectral resolution of ~1 nm. To improve detection sensitivity, the output of several diodes can be combined to achieve sensitivity performance approaching that of the variable-wavelength UV/Vis detectors. Further sensitivity enhancement has stemmed from innovative flow cell design using coated fiber-optics technology to extend the cell pathlength without increasing chromatographic dispersion.

For nonchromophoric analytes (those with no or little UV absorbance), universal detectors such as the refractive index and the evaporative light scattering detectors are typically used. The latter is gradient compatible and is often used in high-throughput screening applications. A new universal detector entry is the corona aerosol discharge (CAD) detector capable of high sensitivity and with the advantage of having similar response factors for most nonvolatile organic analytes.

Mass spectrometry with its excellent sensitivity and specificity is emerging as a premier analytical technique. Two common LC/MS interfaces are electrospray ionization (ESI) and atmospheric pressure chemical ionization (APCI). Common types of MS instruments include the quadrupole, ion trap, triple quadrupole, and time of flight (TOF). LC/MS/MS using triple quadrupole MS is now the primary analytical technique for bioanalytical and many trace analytical assays.

11.6   HPLC SYSTEMS

Several HPLC system trends highlighted below have become evident in recent years.

11.6.1   Low-Dispersion Instruments

Low-dispersion HPLC systems are necessitated by the increasing trend of using shorter and narrower HPLC columns, which are more susceptible to the deleterious effects of extra-column band-broadening. HPLC manufacturers are designing newer analytical HPLC systems with improved instrumental bandwidths compatible with 2-mm i.d. columns by using micro injectors, smaller i.d. connection tubing, and detector flow cells. A new generation of ultra-low dispersion systems dedicated for micro and nano LC is also available.

11.6.2   Ultra-High-Pressure LC

The benchmark upper pressure limit of 6,000 psi has recently been exceeded by a number of research groups (to 200,000 psi) and commercial HPLC systems (9,000–15,000 psi). Higher system operating pressure is the most direct way to increase chromatographic performance by allowing the use of longer columns for higher peak capacities or faster analysis with columns packed with sub-2-μm particles. The key to the success of these ultra-high-pressure HPLC systems is the initial customer acceptance by the research sector and a level of convenience and reliability that is comparable to that of conventional HPLC.

11.6.3   Multi-Dimensional LC

The maximum theoretical peak capacity (n) of a single high-resolution column with ~20,000 plates using gradient elution is about 400—a level insufficient for the analysis of many complex samples. One traditional approach to increase separation power is to use two-dimensional chromatography where the sample is fractionated by the first column and each eluted fraction is subsequently analyzed by a second, orthogonal column. This process can be automated in a 2-D system using switching valves, thus offering a powerful technique for the characterization of very complex protein (proteomics) or peptide mixtures, especially when coupled with time-of-flight MS and bioinformatics software.

11.6.4 Parallel Analysis

Parallel analysis is one of the most effective ways to enhance laboratory productivity. A typical parallel analysis HPLC system consists of multiple pumps or a multi-channel pumping system, a multi-probe autosampler capable of four or eight simultaneous parallel injections from 96-well microplates, and a multiplexed UV detector and/or MS. This generic approach allows a 4 to 8 fold increase of sample throughput while maintaining the traditional performance and convenience of HPLC. This technology can potentially be extended to a 96-channel analysis system.

11.7 LAB-ON-A-CHIP

An exciting area of active research is "lab-on-a-chip," offering the ultimate system in low-cost and high-speed multi-channel analysis. Currently, the technology is successfully applied to CE-based chips in the analysis of biomolecules. Extending this innovative micro-fluidics technology to high-pressure applications in HPLC chips has proven to be much more technically challenging. One notable device is the recent introduction of an HPLC chip integrated with MS inlet probes by Agilent Technologies.

11.8 DATA HANDLING

Chromatography data handling has benefited much from the computer revolution. A PC-based data station typically incorporates method storage, data archival, and report generation as well as full HPLC system control. For large laboratories, a centralized client-server network is becoming the standard system to ensure data security and compliance with regulations. Most network systems also allow the user to access data and to control the system remotely from the office and/or from home via a secured web connection.

11.9 REGULATORY COMPLIANCE

While automated tools and software are available for system qualification, method validation, and system suitability testing, achieving regulatory compliance for an HPLC laboratory remains a complex, costly, and time-consuming process. Many companies are adopting a risk-based approach to balance productivity and compliance. For instance, many pharmaceutical research

Table 11.1. Comparison of an Existing Tablet Assay Method with an Improved "Greener" Method

	Existing method	Improved (greener) method
Solid handling	Grind 10 tablets Transfer all powder into a 500-mL flask	Grind 10 tablets Transfer powder equivalent to ATW into a 100-mL flask
Extraction	Add 400 mL of extraction solvent Sonicate for 15 min QS with extraction solvent Filter the extract	Add 50 mL of extraction solvent Sonicate for 15 min QS with MPA Filter into HPLC vial
Dilution	Pipette 5 mL into a 10-mL flask QS with MPA Transfer aliquot into HPLC vial	
Volume of solvent used/time for procedure	510 mL/45 min	100 mL/30 min

ATW = average tablet weight, QS = fill to volume, extraction solvent = 50% ACN in 0.1 N HCl, MPA = mobile phase A.

facilities are beginning to adopt a two-tier approach, allowing fewer regulatory requirements for early-phase development and strict compliance to regulations for late-stage development and manufacturing.

11.10 GREENER HPLC METHODS

The principles of "green" chemistry by adopting the most efficient and environmentally friendly processes should be practiced whenever possible in the HPLC laboratory. One obvious approach is reduction of solvent consumption by using solvent recycling for isocratic analysis and narrowbore LC columns. Another area is to find ways to reduce sample size and the number of sample preparation steps without sacrificing method performance.[2] A case study to illustrate this principle for a tablet assay is shown in Table 11.1. Another example is an environmental analysis of soil/sediment sample[3] is shown in Table 7.7.

11.11 SUMMARY AND CONCLUSIONS

Table 11.2 summarizes the 10 megatrends in HPLC columns, instrumentation, and applications. HPLC is the premier analytical technique based on mature

Table 11.2. Ten Megatrends in HPLC

1. Shorter and narrower columns packed with smaller particles
2. Novel bonded phases with improved bonding chemistries
3. More reliable pumps with better low-flow performance
4. Higher-sensitivity UV detectors linear to >2 AU; ELSD and CAD detectors for nonchromophoric compounds
5. Faster and more precise autosamplers with less carryover
6. Low dispersion, ultra-high-pressure HPLC systems, multidimensional LC and parallel analysis
7. Lab-on-a-chip microfluidic devices for lower-cost multi-channel analysis
8. Network-based 21 CFR 11 compliant chromatographic data handling with remote access
9. More efficient regulatory compliance with automated tools and software for system qualification and method validation
10. Quicker, more efficient, and greener HPLC methods with fewer sample preparation steps and reduced solvent consumption

technologies that has enjoyed a steady increase in performance and reliability in the past decades. Through competitive pressure, and driven by stringent demands from the pharmaceutical and life science sectors, HPLC will continue to evolve to higher levels of speed, sensitivity, and resolution for many more decades to come.

11.12 REFERENCES

1. J. Naisbitt, *Megatrends: Ten New Directions Transforming Our Lives*, Warner Books, New York, 1988.
2. M.W. Dong, J.X. Duggan, and S. Stefanou, *LC.GC* **11(11)**, 802 (1993).
3. C. Choi and M.W. Dong, in S. Ahuja and M.W. Dong, eds, *Handbook of Pharmaceutical Analysis by HPLC*, Elsevier, Amsterdam, 2005.

INDEX

Absorbance, defined, 88
Absorbance detectors, 87–91
 trends in, 89–91
 UV/Vis, 87–89
Absorption, distribution, metabolism, and excretion (ADME), 137–138
AccQ Tag. *See* Waters AccQ.Tag
Accuracy studies, in method validation, 232
ACD/Method Development software suite, 211t
Acetonitrile (ACN), 28–30
 blending with buffers, 115
Acidic mobile phases, 32
Acidic silanols, 58
Acids
 HPLC analysis of, 162–163
 for mobile phase preparations, 32
Acquity system. *See* Waters Acquity system
Adsorption chromatography, 5–7
Affinity chromatography, 10
Aflatoxins, HPLC analysis of, 167, 168f
Agilent ChemStation graphical user interface, 104f
Alkali metal cations, isocratic separation of, 181f
Amino acids
 HPLC analysis of, 96, 162–163, 185–186
 separation of, 7–9
Aminomethylphosphonic acid (AMPA), 172

Amitriptyline, separation of, 33
Analysis strategies, 123–129. *See also* Pharmaceutical analysis
Analysis systems, dedicated, 102
Analyte information, gathering, 197–198
Analyte peaks, 233f
 in isocratic separations, 205
Analytes
 batch-to-batch reproducibility of, 58
 column selectivity plots for, 42, 44
 pH in the separation of, 31f
 UV spectra of, 199f
Analyte solution, final, 12
Analytical method goals, 196–197
Analytical methods
 limitations of, 12
 need for, 194–195
 range of, 234
Anions, fast gradient separation of, 180f
Antidepressants, retention map and chromatograms of, 31f
Antimicrobial additives, HPLC analysis of, 168–169
API (active pharmaceutical ingredient), determination of, 140. *See also* Multiple API drug products; Two-API drug product
API release, measuring, 148
Application-specific columns, 70
Assay accuracy data, poor, 260
Assays, in pharmaceutical analysis, 139–145
Asymmetry factor (A_s), 24–26

Modern HPLC for Practicing Scientists, by Michael W. Dong
Copyright © 2006 John Wiley & Sons, Inc.

Sensitivity
  enhancing, 132
  of UV/Vis detectors, 88–89
Sensitivity performance, in PDA
  detectors, 93
Separation, analytes in, 11–12
Separation factor (α), 20–21
Separation parameters, fine-tuning,
  204–205
"Sequential isocratic steps" approach,
  202, 203f
Silanols, 53, 54
  types and acidities of, 58
Silica, high-purity, 58
Silica-based columns, 182
Silica-based RPC columns, problem
  areas of, 57
Silica column packing, 53–54
Silica support, increasing surface area of,
  55
Size-exclusion chromatography (SEC),
  8f, 9–10
Small-pore packings, 182
Soft drink sample, HPLC analysis of,
  164f
Software
  design of experiment (DOE), 235, 239
  gel-permeation chromatography, 176
  spectral evaluation, 92
Software for method development, 210,
  211t
Solubility issues, 11
Solute band, longitudinal diffusion of, 37
Solute peak, 17
Solutes, degree of separation between,
  34–35
Solvent degasser, 102
Solvent delivery systems, 81–84
Solvent line filter (sinker), replacing, 244
Solvent preheating, 102
Solvents. *See also* HPLC solvents
  filtration of, 115
  storage and disposal of, 114
Solvent selectivity, effect of, 36f
Solvent strength
  initial and final, 41
  lowering, 205
Sorbents, HPLC, 71–73

Specialty columns, 70
Specificity, in method validation, 230–231
Spectral evaluation software, 92
Split peaks, 253–254
StableBond chemistry, 61f
Stage-appropriate impurity testing, 148
Standard column format, 49–50
Standard operating procedures (SOPs)
  cGMP, 222, 223–224
Sterically hindered group bonding
  chemistry, 60
Sterols, HPLC analysis of, 161
Sugars, HPLC analysis of, 159–160
Summary chromatography report,
  125–126, 127f
Supercritical fluid chromatography
  (SFC), 11
Support type column packing, 52, 53–54
Surface area, of chromatographic
  supports, 54
Synchronous noise, 252
System bandwidth, calculating the effect
  of, 105
System calibration, 196, 227
System contamination, minimizing, 131
System dispersion, 104
  effect on column efficiencies, 106f
System dwell volume, 83–84
System shutdown, best practices for,
  121–122
System suitability samples (SSS), 236
System suitability testing (SST), 224,
  235–237

Tailing factor (T_f), 25, 26–27
Tailing peaks, 254–255
Taste panel method, of capsaicin
  analysis, 166
Thin-layer chromatography (TLC), 11
Time of flight (TOF) MS, 97
Tocopherols, HPLC analysis of, 161f
Toxicity, of organic solvents, 112
Toxicity data, gathering, 197
Trace analysis
  guides on performing, 129–132
  mobile phase in, 130
Trifluoroacetic acid (TFA), 205, 214–215,
  257–258